Nigel Strangeways deals with libel and murder in a London publishing house.

"Admirably solid ... an adroit formal detective puzzle backed up by firm characterization and a knowing, intimate picture of London publishing."
—Anthony Boucher, *New York Times*

"Nicholas Blake's literate and personable detective, Nigel Strangeways, is the best literary creation in the mystery-suspense field since Dorothy L. Sayers introduced her Lord Peter Wimsey over thirty years ago."   —*New York Morning Telegraph*

END of CHAPTER

End of Chapter

by Nicholas Blake

PERENNIAL LIBRARY
Harper & Row, Publishers
New York, Hagerstown,
San Francisco, London

PR
6007
.A97
E55

First PERENNIAL LIBRARY edition published 1977

STANDARD BOOK NUMBER: 06–080397–5

77 78 79 80 5 4 3 2 1

To N.S. *and* I.M.P.
with great affection
from their unsleeping partner
in a very different kind of firm

CONTENTS

I   *Setup*

II   *First Impression*

III   *Stet*

IV   *Verso*

V   *Run On*

VI   *Delete*

VII   *Query*

VIII   *Lower Case*

IX   *Insert*

X   *Revise*

XI   *Bring Back*

XII   *Lead In*

XIII   *Transpose*

XIV   *Wrong Font*

XV   *Close Up*

XVI   *Final Proof*

I *SETUP*

Nigel Strangeways turned into Adelphi, and soon arrived at the distinguished backwater of Angel Street, the Strand traffic roaring softly behind him like a weir. It would be a nice street to live in, he thought—a top-floor flat in one of these tall, elegantly uniform houses: above the hurly-burly, but not cut off from it. As middle age advances and one's youthful illusions recede, almost the only way to get the sensation of starting again, of being reborn, is to move house. But, restless though he was by nature, two moves in twelve months would be carrying one's mobility too far—one must beware of developing a tolerance for that stimulating drug.

His thoughts took Nigel the length of Angel Street, till he reached the high iron gateway which opens on a passage dividing the end of the street from Embankment Gardens. The gardens looked pinched and dismal this raw late-November morning. Beyond them, on the Thames, a tug mooed dispiritedly. Glancing at his watch, Nigel decided to be five minutes early for his appointment, rather than stand about in the cold admiring the apology for nature which these gardens offered.

The office of Wenham & Geraldine occupied the last house to his right, with its frontage on the street and its south side overlooking the gardens. The main door was flanked by two display windows, and there was a smaller door five or six yards up the street. Façade of dark-crimson brick, dazzling white paintwork, exquisite molding and fanlight of the doorway—

1

they created an impression of solidity and grace, charm and decorum, altogether appropriate to the imprint of Wenham & Geraldine. How many Victorian grandees, Nigel reflected, must have carried their bulging personalities up those shallow steps to take a glass of Madeira with the now legendary James Wenham; how many of their ill-conditioned modern successors, surveying this opulent front, must snarl, "So *that's* where the profits go!" A book, the legendary James Wenham had constantly declared, is the precious life blood of a master spirit; and a vast quantity of more or less precious life blood had been siphoned off, over the last century, to swell the tide of Wenham & Geraldine's prosperity.

Moreover, though venerable, the firm was capable of moving with—or at least not more than twenty years behind—the times. One of the windows, Nigel observed, carried a display of the new book by a well-publicized R.A.F. line shooter, together with a model airfield and a collection of toy aircraft dangling from threads above it. The display must be rather wasted on this backwater, he thought; then realized that Angel Street would be a short cut for pedestrians from the Strand to Charing Cross Underground Station, and at certain times of the day busy enough.

He found himself mounting the steps in front of a well-preserved lady, whose equine face made her choker of pearls resemble a horse collar. He opened the door of the reception room for her, and stood aside. She swept past him, with no other acknowledgment of his politeness than a puff of scent, and disappeared through a door at the far end of the room.

"My name is Strangeways," he told the receptionist. "I have an appointment with Mr. Geraldine."

After applying herself to the house telephone, the girl announced, "If you will take a seat for a moment, Mr. Geraldine's secretary will come down for you."

Nigel was getting the Number Two treatment, as later he would learn. The Number One, reserved for the equivalent of Royalty among the firm's writers, was for a partner to come downstairs in person. Lesser V.I.P.'s were fetched by a secretary; while the rabble had to find their own way. Nigel filled in time now by studying the signed photographs of authors which adorned one wall of the reception room. The earlier ones ran to beards and expressions of quite unnerving self-confidence; as the eye moved on toward present times, the faces gradually lost both hair and assurance, the most recent being marked either by grinding *Angst* or by that pop-eyed, implausible bravado which looks out so often from the photograph files of Scotland Yard. To this, however, there were two noticeable exceptions: a heavily-mustached type in military uniform—the photo was signed "Richard Thoresby"—and the lady who had brushed past Nigel at the door. The large signature on her photograph informed him that she was none other than Millicent Miles—the glamorous, the unspeakable Millicent Miles, that queen of best-sellers who changed her publishers almost as often as once she had changed (so rumor said) her lovers. It was odd that Wenham & Geraldine should have taken her on. Well, here she was—full evening dress, tiara, perfectly waved hair—gazing haughtily from the wall.

"Wasn't that Miss Miles who came in just before me?" asked Nigel.

"Yes. She is working here," the receptionist replied in neutral tones.

"In the firm, you mean?"

"Oh no. She's writing her memoirs for us: and when she was moving house, Mr. Geraldine put a room here at her disposal."

"Memoirs? That should cause a stir."

"We hope to do well with them," said the girl demurely. "Of course Miss Miles is not the type of author we often have on our list."

"I should say not! Like a university press publishing Elinor Glyn."

Nigel studied the girl. A handsome creature, in a gipsyish way; about twenty-three and tries to look older; very self-contained behind her horn-rimmed glasses; couldn't have worked here long, but was already using the publisher's "we" as to the manner born. Some quality in her speech made him ask,

"Were you at Somerville?"

"Oh dear, does it stick out like that?"

"What did you read?"

"History."

"Do all right?"

"Well, actually I got a First." The admission, accompanied by a gauche sideways jerk of the head, took several years off her apparent age.

"But you're stuck at the end of a telephone?"

"The firm likes one to start at the bottom of the ladder. If I make good with the callers, they'll promote me to secretary—and some reading, perhaps."

"The Victorian regime, eh? I see you're writing a book in between times."

Flushing, the girl thrust some manuscript pages under a blotter. "You see too much. Sorry, but I hate my work being overlooked."

"Most writers do, I believe—while it's in progress. No doubt the anthropologists could tell us something about that."

At this point another studious-looking girl appeared, Mr. Geraldine's secretary. Nigel followed her through the inner door into a passage which led, he later discovered, from the smaller street door to the packers' room. One side of this passage was heaped with bulging sacks; on the other was a lift, whose outer doors struck Nigel a vicious slap from behind as he entered.

"They always do that," said the girl brightly, as if explaining the foibles of a favorite but unreliable domestic animal.

"Even to Millicent Miles?"

The girl giggled. "Well, I believe she *did* ask Mr. Geraldine to have a new lift installed," she said; then gave Nigel a startled look, which plainly inquired how she could have let fall such an indiscretion before a total stranger—a question a good many people, who came across Nigel in his professional capacity, had cause to ask themselves.

Reaching the second floor, Nigel was led to the senior partner's room. A large, high, rectangular room, lit by a Venetian glass chandelier transformed for electricity and two sash windows overlooking the street: the wall facing the windows lined with books; white paneling; a luxurious fitted carpet. Nigel's arrival caught its three occupants in what might have been the *dégagé* and eternally-arrested pose of a modern conversation piece—the bald-headed man sitting at his desk flanked by the woman leaning negligently against the further window, and the saturnine red-haired chap gazing down at his shoes and jingling coins in a trouser pocket. For a second or two the poses were held. Then the bald man rose and advanced smoothly upon Nigel.

"Mr. Strangeways? Exceedingly good of you to come along. May I introduce you to my partners? Miss Wenham—the granddaughter of our founder. Mr. Ryle, who has recently joined us."

Arthur Geraldine had the portly figure, the bland face and address which convention associates with bishops and butlers; he had, also, a surprisingly muscular hand grip.

"You will take a glass of Madeira with us? We keep up the old—"

"Filthy stuff. Like liquid demerara sugar," Miss Wenham broke in.

"My dear Liz, you are incorrigible."

They sat at the mahogany table in the middle of the room. Arthur Geraldine poured out Madeira for the men; Liz Wenham treated herself to a glass of lemonade from the bottle on the silver tray, and bit noisily into an apple. A dumpy woman of about fifty, gray-haired, with very white teeth, rosy-russet cheeks, and a mild, bright, clear eye, she would have looked more at home, Nigel felt, in tweeds, in a cottage-weaving establishment in Westmoreland. For all her fresh-air appear-

ance, though, she soon revealed a businesslike side, cutting in
on Arthur Geraldine's civilities with

"Well, hadn't we better get down to it? Joan, take Mr.
Geraldine's calls for the next half hour."

The secretary whisked out of the room like a leaf in a
draft. Basil Ryle stopped chinking the coins in his pocket. A
twinkle, whether frosty or humorous Nigel could not be sure,
appeared in the senior partner's eye.

"Quite right, Liz," he said. "Miss Wenham keeps us all on
our toes. Now, let me put our little difficulty before you,
Strangeways. It is understood, of course, that this is highly
confidential?"

Nigel nodded.

"You will be aware, perhaps, that this firm specializes in
memoirs, biographies and the like. It would be no exaggeration
to say that the Wenham & Geraldine imprint on any volume
of this nature is a guarantee of the highest quality."

Mr. Geraldine's flowing periods were interrupted by a loud
crash as Liz Wenham's teeth took another segment out of
her apple.

"Some two years ago we entered into an agreement with
General Richard Thoresby to publish his autobiography.
When we received the manuscript, early this year, we dis-
covered that parts of it were—ah—extremely controversial."

"Riddled with libel," Miss Wenham mumbled through her
mouthful of apple.

"In particular, where the author criticizes the conduct of
operations, during the last war, by Major General Sir Charles
Blair-Chatterley. Chatterley, it seems, is his bête-noire—"

"The chap who was recently Governor-General of—?"

"The same. And Thoresby had also made the most serious
allegation about Blair-Chatterley's handling of the disturbances in the colony in 1947. I will not trouble you, for the
moment, with details. Suffice it to say that we made repre-
sentations to the author, pointing out the relevant passages,
and discussing with him how they should be deleted, or at

least toned down, to avoid the danger of libel. He proved somewhat intransigent—"

"A blinding nuisance," Liz Wenham put in.

"—but we finally reached, or thought we had reached, agreement with him."

"I take it you had submitted the manuscript to your legal advisers?" asked Nigel.

"Naturally."

"It sent them right up the wall," Miss Wenham remarked.

"However, when we received proofs, last July, we discovered to our alarm that the author had failed to delete from the manuscript two of the most offensive passages, which he had previously agreed to take out."

"But hadn't the manuscript been examined again before you sent it to the printers?"

"There was a slip-up there. Ryle, who was looking after the book, had gone on holiday; and in his absence, Protheroe, our chief reader, simply marked it up for the printers—it's a standard format—and handed it to our Production Manager to send off."

"We were badly behind schedule with the book, and things got rushed at that point," Liz Wenham explained. "We went straight into page proof so as to get it out for the Christmas market."

"I see. Right. It's July. You've got your page proofs and found two of the offending passages still there. What next?"

"We asked General Thoresby along," said Mr. Geraldine, "and pointed out that we couldn't go to press unless the passages were revised or deleted. Our libel lawyer was present at the conference. It was quite a sticky one."

"The General behaved like an overgrown schoolboy," said Miss Wenham.

"He doesn't know the difference between a pen and a sword." Basil Ryle spoke for the first time—a nasal but not unpleasing voice; his eyes had been moving steadily from one speaker to another during this conversation.

"Well, the long and the short of it is that the General lost his temper and deleted the passages, rather violently, in our presence," said Geraldine.

"The word 'emasculated' was heard to fall from the warrior's lips," Basil Ryle murmured.

"Fortunately no serious repagination was involved. One of the passages only ran to a few lines, and the longer one came at the end of a chapter."

"No repagination was necessary, anyhow," said Liz Wenham brusquely. "Someone stetted the two deletions."

"Miss Wenham means," explained Arthur Geraldine, "that during the period which elapsed between our conference with Thoresby and the sending the book to press, someone marked the deleted passages with dots and the word 'stet'—in block capitals, so that the handwriting can't be identified. 'Stet' is the sign to a printer that the marked passage be retained."

"Strangeways knows all about that." Liz Wenham's voice crackled with impatience.

"Of course, of course. I was forgetting that you had put out a book or two yourself. And very charming work too, if I may say so."

Nigel felt like bearings bathed in oil, under the mellifluous tones of Mr. Geraldine. He bowed, and caught a quizzical glance from Miss Wenham.

"Where was I? Ah yes. Our representatives traveled proof copies—with the offensive passages heavily inked out, of course—and reported a most satisfactory response. We raised the first printing to fifty thousand. The book came out two days ago. Yesterday we had a communication from Sir Charles Blair-Chatterley's solicitors. They have obtained an interim injunction, and it is clear that their client will set his face against any settlement out of court. We took steps instantly to have the book withdrawn, but the damage has already been done."

"Just a minute," said Nigel. "Presumably there were advance copies. And nobody here noticed that the libelous passages had been put back? I'd have thought that some at least of the reviewers would have noticed too, and tipped you off before publication day."

"I'm afraid the responsibility is mine," began Basil Ryle, with something less than his normal self-assurance, after a few moments' embarrassed silence.

"Nonsense, my dear boy. You mustn't reproach yourself." Mr. Geraldine turned to Nigel. "I should explain that each title we publish is given a general supervision, from start to finish, by one of the partners. Ryle was in charge of Thoresby's book, certainly. But, once final proofs had been passed, there was no onus upon Ryle to scrutinize the text of the advance copy. Nor, in fact, would he have had time to do so. He looks after all our advertising and publicity work, and he's been heavily engaged with the promotion campaign for this title in particular."

Nigel was aware of a certain tension in the room during this apologia: it seemed to emanate from Liz Wenham, whose nails were beating a tattoo on the table.

"What about the reviewers?" he asked.

"Owing to the delays caused by the author's, hm, attitude, we were badly behind schedule. Copies only went out a week before publication date—to postpone that date would have been inadvisable, for several reasons. I can only suppose that the reviewers had not had time to—"

"Or the will to," put in Basil Ryle. "Blair-Chatterley is far from being universally loved by the military profession, and *Time to Fight* is a book that would be given to military experts for review."

"It's news to me that reviewers read books." Miss Wenham sounded impatient. "Anyway, the point is that we're in the doghouse."

"You're insured against libel, I presume?"

"There are no insurance policies against loss of prestige," replied the senior partner in oracular tones. "Technically, as you know, publisher, printer and author are held equally responsible in an action for libel. Even supposing our insurance policy will cover our share of the damages—and they're likely to be enormous in the present case—there's our reputation to consider. After such a flagrant libel, authors and agents might well fight shy of sending us their books: our whole reputation for—" Arthur Geraldine's pink cheeks quivered, his voice shook. "Why? I simply can't understand it. We've always been a happy family here. Why should anyone want to do such a thing?"

It was a rhetorical question; but Nigel had little tolerance for rhetorical questions; he proceeded to answer it, dogmatically.

"There are four possible motives. To damage your firm. To do down the author. To disoblige this Blair-Chatterley. Or, of course, pure mischief. I take it you want me to find out—"

Arthur Geraldine held up his hand: there was more hair on its back than on his head. "Precisely. We are not vindictive. But we must prevent any recurrence of the trouble. Finding the person responsible won't help us in court; but we should at least feel secure against—"

"Suspect anyone?"

Automatically, the three partners' eyes sought one another in a silent consultation. Nigel could read their minds: they were debating the ethics of voicing suspicions for which they had inadequate grounds, or which they were ashamed of feeling. After a short pause, Mr. Geraldine spoke:

"We have discussed it amongst ourselves, naturally. But we are quite at a loss. During the crucial periods, the proof copy of *Time to Fight* was in Protheroe's room, accessible to—"

"I'd rather go into the mechanics of it later, if you don't mind. There was no one with a grudge against the firm?"

A slightly embarrassed silence, broken by Basil Ryle: "I still think Bates should be questioned."

"We've had all that out, Basil." Miss Wenham was visibly bristling.

"Bates is—was our Production Manager," Arthur Geraldine explained. "He was certainly under notice at the time. But he'd been with us, let me see, thirty-two years, and always devoted to the firm's interests. I am confident he is above suspicion."

"But you sacked him."

Mr. Geraldine winced at the crude expression. "He was getting on in years, you know, and his methods had become just a leetle—ah hum—over-conservative, shall we say? So we asked him if he was prepared to retire, somewhat before the fixed date, on full pension of course."

This Agag-like treatment of the subject only heightened the tension round the table. Nigel interpreted it—correctly, as it proved—to mean that the up-and-coming young Ryle had fallen foul of Bates or his methods, and forced his resignation, against the opposition of Liz Wenham certainly and perhaps of the senior partner.

"And was he?"

"Was he—?"

"Was your Mr. Bates 'prepared to retire'? Did he go, or was he pushed?"

Nigel caught an approving gleam in Miss Wenham's eye, as she said, "He certainly did *not* want to go. But he wouldn't do a thing like that. It's out of the question."

Basil Ryle, sighing, shrugged his shoulders. "I think, if Mr. Strangeways agrees to help us, he must be allowed to have an open mind about everyone in the firm—including us round the table here."

"Ah, come now, Basil." A trace of Irish was heard, for the first time, in Arthur Geraldine's speech.

"Basil is right. We owe it to our employees." As so often when Liz Wenham spoke, it had the effect of a window opened to let in some bracing air. The uneasiness in the room was dispelled.

"What about your reader?" Nigel asked. "By the way, is he the Stephen Protheroe who wrote *Fire and Ash?*"

"Yes. A remarkable poem, wasn't it?" said Miss Wenham.

"We have sold some twenty-six thousand copies of it," remarked Mr. Geraldine, "since 1927, and there is still a modest demand. Extraordinary, you know, that he never followed it up."

"I should think that poem would be enough to burn anyone out. And is he to be above suspicion too?"

"Stephen has been with us since 1930. He has no financial interest in the firm, though we did sound him about partnership a few years ago."

"He seems to have had the best opportunity to tamper with the proof copy," said Nigel.

"Possibly." Arthur Geraldine's very long, thin upper lip seemed to stretch yet further, giving him a sharklike look. "But if he did it, I'd never trust my own judgment again."

They discussed ways and means for a while. It was decided that Nigel should pursue his investigations, under cover of being temporarily engaged as a specialist reader. None of them supposed that this cover would hold up for more than a day or two, but even so short a period might help. His fee was stated, and agreed to. Then Miss Wenham said,

"Where's he going to work?"

"Better put him in the room next to Stephen's," said Arthur Geraldine.

"Miss Miles won't welcome that," said Basil Ryle.

"Oh damn, I'd forgotten she was still with us. How much longer do we have to entertain the woman?"

Ryle shrugged. It was evident to Nigel that the best-selling authoress was Ryle's protégée, acceptable to the other two, if at all, for the money Wenham & Geraldine would make out of her.

"How long has she been using the room?" he mildly inquired.

"Since we signed her up, last June, off and on. She was moving house then."

"And now she seems to have moved in permanently on us," muttered Liz Wenham.

Geraldine said, "Well, as long as it's a golden egg, I suppose we can put up with her laying it on the premises. Mr. Strangeways had better go in with Stephen, then." Moving over to his desk, he dialed a number on the internal telephone.

"Stephen, could you spare us a few minutes?"

The receiver buzzed at him like a wasp.

". . . Well, it is rather important. Thanks." He put down the telephone. "Stephen's in a mood."

Nigel looked forward with some interest to meeting the author of that acrid, tragic poem, *Fire and Ash*, in a mood or out of it. Stephen Protheroe, once upon a time "a name to be conjured with," "a young singer to be watched," etc., etc., had given the public no further opportunity to watch him or to conjure with him. Withdrawing again into the total obscurity from which *Fire and Ash*, his one published work, had emerged, he had been lost to view for over twenty-five years. Few of the many admirers of that book could have stated with any certainty whether he was still alive, or dead.

"We're all very fond of Stephen," said Liz Wenham. "And terrified of him."

11 *FIRST IMPRESSION*

A little grig of a man darted into the room, with a movement that suggested fins rather than legs. The fishlike effect was accentuated by his thin mouth, which was opening and shutting silently. After a few seconds, the mouth achieved speech.

"How much longer is that bloody woman going to infest the building? She's wasted half an hour of my time this morning already—pestering me about her punctuation. Punctuation! I ask you! Am I hired to put in stops for female morons? And I don't like her scent either."

Protheroe's diatribe was itself most effectively punctuated —with loud sniffs. Nigel took in the high-bridged nose, the noble forehead, the fine eyes blazing through horn-rimmed spectacles, which surmounted so oddly a weak mouth and an almost nonexistent chin. The man's voice was equally paradoxical—deep and resonant, but occasionally cracking to a furious squawk. One's eyes kept returning to his mouth, which had the habit of nibbling silently at words, before pronouncing them aloud.

"I have worked in this firm now for a quarter of a century," Protheroe was proclaiming, "and I should have thought I had earned the right to a little (sniff) privacy. Of course, if you prefer me to spend my time spoon-feeding illiterate (sniff) ex-trollops rather than reading manuscripts, I'll accommodate myself to your (sniff) new policy. Who's this?"

Stephen Protheroe had at last become aware of Nigel's

14

presence, and whipped off his spectacles with the apparent object of seeing him better.

"Our new reader. Mr. Strangeways—Mr. Protheroe."

"Hm'ff. Sorely needed. How do you do? You can take Miss Millicent Miles off my hands. I like the 'Miss.' A Miss is as good as a Miles, I told her just now. She always—she hates puns. So do I, but it's the only way I can get her out of the room." He turned upon Liz Wenham. "New reader, did you say? We don't want a new reader."

"Now sit down, Stephen, have a glass of Madeira, and calm yourself."

Protheroe peered at the proffered glass suspiciously, took a gulp, smacked his lips.

"The Madeira vine," he informed the company, "was introduced to the island by the Portuguese, from Cyprus, or possibly Crete, early in the fifteenth century. It is a cousin of the wine known to schoolboys as Malmsey, and to antiquarian pedants as Malvoisie. Under either name it no doubt tasted equally sickly."

"Stephen, could you stop talking for a minute and listen," implored Liz Wenham. The note of exasperated affection in her voice was not lost upon Nigel.

"Strangeways has very kindly consented to investigate our little trouble over *Time to Fight*. You remember, Stephen, he was recommended to us yesterday by—"

"Nobody here ever tells me anything."

"You just don't listen—that's your trouble," said Liz Wenham.

"I shall observe your methods with interest." Protheroe gave Nigel a courtly bow. "The science, or perhaps one should say the art, of detection has long been to me—"

"Perhaps we should let Strangeways proceed with his art, or science?" Basil Ryle interposed.

"'In my craft or sullen art,'" chanted Protheroe. "A shade dubious, do you think? That 'sullen'—did it come from the heart or the head?"

"For God's sake, Stephen!" exclaimed Arthur Geraldine. "This is a publishing house, not a Senior Common Room. You don't seem to realize the gravity of this libel action."

"Protheroe has nothing at stake," snapped Ryle.

Liz Wenham bristled. "I think we all know how much spiritual capital Stephen has put into the firm."

"Now, children! No quarreling!" Protheroe turned to Nigel. "At your service."

"I'd like to study the proof copy that was tampered with, and the original typescript. I want the addresses of General Thoresby, Blair-Chatterley and Mr. Bates, and a list of your present employees, with the length of time each has been in the firm, and a specimen of his or her handwriting. In the meantime, perhaps I could have a talk with Protheroe."

Arthur Geraldine nodded. He went to a corner of the room, opened a safe which was discreetly concealed behind a large screen—as though nothing to be associated with base lucre should meet the eye in this high-toned room—and returned to the table with a proof copy and a typescript bound in brown paper.

"Er, you will be at pains, I am sure, my dear fellow, not to—" The senior partner, looking oddly pathetic, broke off and tried again. "I do realize, of course, that you will have to ask a great number of questions. But we do not want the staff, er, well, you know, unduly—I mean, unnecessarily—upset."

"I am sure Mr. Strangeways is the soul of tact," said Liz Wenham soothingly.

With that, Nigel took his leave, escorted by Stephen Protheroe. Stephen pointed out the rooms of the other two partners, off the same passage as Geraldine's. Then they mounted a flight of stairs, emerged onto a landing with the lift doors to their left and a notice, WENHAM & GERALDINE: EDITORIAL DEPT., on the wall in front of them, and stopped at a door three yards down the passage. To this door was fastened a card, bearing in large block capitals the legend MR. BLODWORTH—KEEP OUT.

"Here we are," said Stephen. "Mr. Blodworth passed over many years ago, but we like to keep in touch with our past. And of course it does baffle a certain number of my unwelcome visitors—though not, I fear, for long." He gazed at the card meditatively; then, whipping out a thick pencil, wrote underneath KEEP OUT the words THIS INCLUDES YOU, MISS MILES. "Welcome to my hutch," he said, throwing open the door.

It was certainly very different from the great open spaces of the senior partner's room.

"Very cozy, I am sure," said Nigel, eying with misgivings the cluttered desk, the naked bulb overhead, the shelves of dusty books, the two hard chairs which comprised the furniture and fittings of this glorified packing case. One side of it was partitioned off from the room next door by a thin wall with a sliding, frosted-glass window in it. Through the wall could be heard the excitable chatter of a typewriter.

"She actually *composes* her sickening works on a *machine*," said Protheroe, with an expression of deep disgust. "That shows you, doesn't it?"

"And where do I sit?"

"Oh, I expect we can squeeze a little table into that corner. If we do, it will be physically impossible for our neighbor to get into the room at all. Then it only remains to seal up the sliding window and we shall be inviolable."

"Yes, that's very nice. But how about getting *out* of the room? We shall need to do that from time to time."

"Oh dear me, are you one of those *active* detectives? I was hoping you would be the Nero Wolfe type, never budging from your chair."

Clearing a space on the desk, Stephen put down the proof copy and typescript of *Time to Fight;* but Nigel showed no eagerness to examine them yet. Instead, he flicked a forefinger at them and asked, "Well, what can you tell me about all this?"

"Tell you? How do you mean?" Protheroe seemed a little disconcerted.

"Go back to last July. The day this proof came back from the author. What date was it, by the way?"

"The 22nd."

"Did it come by post, or did the author bring it in?"

"Post. Does that matter?"

"Probably not. What happened next? What's the exact procedure in this house?"

While the typewriter clacked feverishly next door, Nigel extracted from Stephen Protheroe the following information. The proof had arrived by midday post on the 22nd. The parcel was brought to Basil Ryle's room. After lunch he began scrutinizing it, to make sure that the author had settled all the printer's queries, and the author's corrections made sense. A publicity meeting that afternoon prevented Ryle from finishing work on the proof, so he took it home with him. Reading it after dinner, he discovered that the offending passages had not been taken out. Next morning he discussed with Geraldine and Protheroe whether they should be quietly deleted in the office and the book sent to press, or whether General Thoresby should be consulted. Though the General had previously agreed to their deletion, Geraldine decided that the firm had better have a showdown with him, and get the record absolutely straight. So a meeting was arranged for 2 P.M. the same afternoon, between the partners, their legal adviser, and the author. This, as Nigel had already been told, was a lively encounter; but the General had finally been induced to see reason, and the libelous passages were marked out there and then. The meeting broke up a little before 3 P.M. Ryle at once brought up the proof copy to Stephen Protheroe, asking him to examine it for any errors which might have gone unnoticed by printer's reader and author.

"Is that part of your job normally?" asked Nigel.

"Oh, I'm just the little old dogsbody who sits up aloft. Ryle would have done it, but he was immersed in one of his publicity campaigns. He's a (sniff) live wire, you know. High-voltage fellow."

"So now we come to the zero hour. About 3 P.M. on July 23rd. You're correcting the proof. How long did it take?"

"I'd got most of it done by 6:30—I'm pretty fast, you know, and a Deadeye Dick for misprints. I finished it off next morning, by about 11, and gave it to Bates to post."

"What part of the book do the libelous passages come in?"

"Chapter III and at the end of Chapter XIII."

"And had you got past Chapter XIII before you left that evening?"

"Yes. I'd done fourteen out of the seventeen chapters."

"So the proof must have been tampered with *after* you left, or the next morning? Otherwise you'd have noticed the stet marks while you were reading it through that afternoon?"

"I certainly would."

"You're sure of that?"

"Positive."

"Well, let's take the evening first. Have you any idea who was in the building after you left?"

The partners had discussed this with him yesterday, Stephen said. It was impossible to be certain even of their own movements on a night over four months ago. But, as far as they could remember, Geraldine had gone up to his flat on the top floor shortly after 6:30, Liz Wenham had left the office about 6:15, Basil Ryle had worked till 7 and then departed. As for the staff, some of them normally finished work at 5, the rest at 5:30, the only exceptions being the partners' secretaries, who on occasion did overtime. The records showed that, on the night of the 23rd, none of these secretaries was kept late.

Further questions produced for Nigel the information that the main door was locked and bolted at 5:30; anyone leaving after 5:30 would go out by the side door. The receptionist remained at her post till 5:30, and had already told the partners that General Thoresby had not returned to the building after his departure.

"What about this side door? Who has keys to it?"

"The partners. Myself. And there's a spare in the receptionist's desk. I believe Millicent Miles has borrowed it once or twice when she wanted to come back and work late."

"So, after 7 that evening, there was no one in the building, officially, except Mr. Geraldine?"

"And his wife. As far as we can tell."

"What about Miss Miles? Was she here that day?"

"I simply can't remember. It was so long ago."

Nigel threw his head back and gazed down his nose at Stephen Protheroe.

"Perhaps it strikes you as odd that I want to know about the movements of the people who should be most above suspicion. But who else—except the partners, the author and yourself—would have known, at the time, about the libelous passages?"

Stephen gave a snuffling laugh. "My dear chap, you've no idea the gossip that goes on—"

"But how could it *start*? From whom?"

"Jane, for example. Arthur's secretary. She was taking notes at the meeting with Thoresby that afternoon."

"Yes, I've noticed she is not perfectly discreet."

"You've got to realize that in a publishing house—odd as it may seem—a lot of people are genuinely interested in books; anyone attached to the partners or the editorial staff almost certainly is. I bet you the news about the General's trying to put one over us was all round the building in a few hours."

"I'll believe you. Next morning, then. You arrived at—?"

"Nine-thirty."

"The page proofs were where you had left them."

"Not exactly. But the cleaners always shift everything on my desk."

"You didn't casually turn back to the deleted passages?"

"No. I went straight on correcting from where I had left off."

"Were you in this room the whole time from 9:30 till 11 when you took the proof copy to Mr. Bates?"

"I've been trying to remember, ever since this thing blew up. I was certainly in the loo for five minutes, being a man of regular habits—probably at about 9:45. And of course one pops out now and then to have a chat with somebody. It's really impossible to be accurate at such a distance of time."

"So, theoretically, there were opportunities for the proof to be tampered with any time between 6:30 P.M. on the 23rd and 11 A.M. on the 24th: after that, only Bates himself or some Black Hand at the printer's could have done it.

"You can dismiss Bates from your mind at once. He's an atrocious old bore, but he'd never do anything to injure the firm. But all this is assuming that I've been telling you the truth," added Stephen with an impish look.

"We'll work on that assumption," Nigel answered. "For the present."

"The trouble is, I had so much more opportunity than anyone else."

"What would be your motive, then?"

"Ah, there you have me. I really can't imagine."

Nigel's pale blue eyes were alive with interest—an infectious though impersonal interest, which had drawn confidences, almost hypnotically, from many people whose interests had by no means been served by making them. He leaned back and said, "Well, who do you think did it?"

Stephen's thin lips, turned down at the corners, nibbled for words. What he would have replied Nigel was not to know; for the sliding window behind Stephen's back opened, and a voice said, "How do you spell 'holocaust'?"

"It's a word I never use," Stephen sourly answered, over his shoulder.

"Haven't you got a dictionary there?" Millicent Miles' head protruded through the window, like a horse's looking over the top of a loose box. "Oh, you've a visitor."

"Yes."

"Won't you introduce me?" asked Nigel, rising to his feet.

"Mr. Strangeways—Miss Miles." Stephen added grumpily, "He's come to do some reading for the firm."

Millicent Miles extended a jeweled hand through the window. "I'm so glad. I hope you'll be on my side. You must talk Mr. Protheroe round: he's so obstinate."

Nigel had no idea what she was talking about. Pinched Knightsbridge accent; black woolen dress, a little dandruff on the shoulders; the pearl choker; large mouth and prominent teeth; green, rather insolent eyes, with that wandering look to be found in the eyes of authors and snob hostesses. So much, Nigel took in at once. Conversation through the sliding window was difficult: one had to stoop sideways, as if talking to a booking clerk.

"How's the book getting on, Miss Miles?" he asked.

Her eyes rolled up and round—a trick of hers, he was to learn, and oddly gauche in a woman so self-possessed; it must be a throwback, he thought, to the days when she was young, shy, intense, awkward.

"It's a dreadful struggle," she said.

"I thought your births were always painless," remarked Stephen Protheroe.

Miss Miles laughed, showing her large teeth. "That's a common masculine delusion, isn't it, Mr. Strangeways?"

"Not in this case," said Stephen. "Your typewriter has never paused for breath this morning. You write in a condition of self-hypnosis, you know."

"I do *write*, anyway," she retorted, smiling too sweetly at Stephen's averted face.

These exchanges puzzled Nigel. They sounded like the wrangling of an old married couple, the weapons' edges blunted by long use. Protheroe, of course, was an eccentric, a law unto himself: one did not expect from him the normal amenities and evasions of social intercourse.

"I suppose you've known each other a long time?" he asked.

The other two spoke almost together.

"It seems a very long time."

"We've been neighbors since June, when I began using this room. I've got quite used to Mr. Protheroe's idiosyncrasies. His bark is worse than his bite."

Nigel smiled. "For me, as Christopher Fry would say, the Bark is Bite Enough."

Miss Miles gave her loud, rattling laugh. "I must remember that."

"She'll put it into her book," muttered Stephen, "with or without acknowledgments."

"Come and talk to me, won't you, Mr. Strangeways. I've done my morning's stint. And I really can't converse doubled-up like this."

Nigel raised his eyebrows at Stephen, who seemed sunk in a pit of disgust; then, picking up the proof of *Time to Fight*, he went out into the passage. The door of Miss Miles' room, he noticed, had a new lock. The room was larger than Stephen's, but sparsely furnished. A table and office chair stood on a rug in the middle, the chair's back to the door. An armchair, an electric fire, a vase of flowers on the window sill, and a typewriter: little else. A neat pile of typescript was laid beside the machine. Pinned to a wall, a piece of squared paper caught Nigel's eye.

"That's my work graph," said Miss Miles. "Every day I extend the line to show how many words I have written. It keeps me up to the mark."

"Very businesslike of you."

"Well, writing is a business. I've no patience with the highbrow attitude. I have a certain commodity at my disposal. I wish to sell it in the best market. I've got to keep up my output." She rolled her eyes. "Are you very shocked?"

Nigel made a polite, deprecatory sound.

"How long have you been in this trade? I don't seem to remember—"

"I do a bit of specialized reading from time to time."

"I think specialization is the curse of our age." Miss Miles delivered the truism with all the animation of one who has planted the flag on virgin soil. "Hallo," she went on. "You've got Thor's book."

"You know him well?"

"I know too many people. It's the bane of a successful writer's life—one simply can't get away from one's public. Lectures, cocktail parties, fan mail, press interviews—sometimes I wish I'd never set pen to paper." Miss Miles sighed, rather theatrically.

"But you enjoy it? And it's all grist to the autobiography."

The popular authoress gave him a little-girl look. "Do you agree with me that one's autobiography should be absolutely, fearlessly frank?"

"Well, about oneself perhaps. But if you start lacing into other people—" Nigel held up the proof copy.

"Oh yes. I believe there's been some trouble about Thor's—"

"You know damned well there has," came a barbed voice from the next room. "And exactly what trouble too."

Millicent Miles rose from her armchair and shut the sliding window.

"Odious little man," she muttered. "But he's a genius in his own line. One must make allowances."

In talking with an egotist it is never difficult to steer the conversation into a given channel without his perceiving it: the difficulty is to keep it there. Millicent Miles' egotism, Nigel judged, had much in common with a child's; and no doubt it was the permanently non-adult part of her mind which had made her so prolific and successful a writer of romances. At any rate, she showed no reluctance to talk about the libel imbroglio, and no suspicion that Nigel was anything but a personable stranger who enjoyed conversing with Millicent Miles. Nigel was careful, of course, to keep her at the center of the picture. A novelist such as she, he disingenuously put forward, had special powers of observation and character judgment. Were he one of the partners, it was she to whom he would come first for advice, particularly as she'd been working next door to the room where the proof copy had been tampered with. Naturally, she must have been concentrating upon her own work at the time; but often the subconscious, particularly of so sensitive a person as Miss Miles, took in things its owner was unaware of till later. But perhaps she had not been in the office at all on July 23rd and 24th?

The question was quickly settled by a glance at the wall chart. The graph line was solid for the days she had worked in the office, dotted for those when she had been absent: the line for July 23–24 was solid. But, as to any furtive comings-and-goings in the next room on those dates, her mind was a blank. People were constantly popping in and out of one another's rooms here, she complained; that was why she had asked Mr. Geraldine to have a lock put on her door.

Since Millicent Miles' interest in the affair of the proof copy was confined to her own presence near the scene of the crime, Nigel could pursue the subject no further—not, at any rate, in his role of temporary reader. He tried another opening.

"When you came into Protheroe's room just now, you said something about hoping I'd be on your side."

"Yes, that's what I wanted to talk to you about, privately." She lowered her voice, glancing at the partition between her room and Protheroe's. "When Mr. Ryle asked me to give this firm my autobiography, there was an understanding that Wenham & Geraldine should also reprint some of my novels—the earlier ones which have been out of print for some time. I am convinced—" Her green eyes took on a faraway look, as though she were reverently contemplating some spiritual revelation—"I am convinced that they have a message for the younger generation today. Which makes Stephen Protheroe's opposition all the more disgraceful."

"You say there was 'an understanding' between you and Mr. Ryle. But you've no clause in your contract—?"

"It was a gentleman's agreement. One expects such understandings to be honored by any reputable firm."

"But surely Protheroe hasn't all that much influence here?"

"Oh, he's their little tin god"—a certain coarseness of intonation began to show through her Knightsbridge accent—"and Miss Wenham for some reason has taken against me. Ridiculous frump! Between the two of them, they can put such pressure on Arthur Geraldine—he always was a weak man—"

Nigel let her rattle on, repeating her grievance with every variation known to female ingenuity. Preserving (he hoped) the respectful but noncommittal expression of a new employee of the firm, who must neither be disloyal to its policy nor discouraging to one of its lucrative authors, Nigel, under cover of the typewriter, idly riffled through the pages of *Time to Fight* till he found the chapter ending which had caused the trouble. He saw a stet mark against the last paragraph. Then a word in it rose up at him—the word "holocaust."

When Millicent Miles had talked herself to a temporary standstill, Nigel rose to go, assuring her that he would do everything within his very limited power, etc. etc. As he went, his hand brushed against the pile of typescript lying face down on the table, and dislodged some of the top sheets. Apologizing profusely, he picked them up from the floor and replaced them—but not before his eye had swiftly taken in the last sentences Miss Miles had typed before opening the window into Protheroe's room. The word "holocaust" did not appear in them; nor, even allowing for the wild vagaries of her style, did they offer any conceivable context for such a word.

III *STET*

Half an hour later, Nigel Strangeways was eating sand-
wiches and drinking Scotch ale in a pub off the Strand. The
place was patronized chiefly, it seemed, by the lower ranks of
business and the civil service, the latter distinguished by
large, black Homburg hats crammed down over their ears and
a self-important manner of speech, the former by a drab and
synthetic bonhomie which might have been learned from a
chapter on the man-to-man approach in some manual for
commercial travelers. The two types were identical, however,
in one respect—that hideous and slovenly accent which is the
marriage of dying Cockney with a degraded culture. No wonder, thought Nigel, people devour the impossible romances of
Millicent Miles: reality is altogether too sordid. But reality, if
one looked round at a place like this, was also quite unreal.
These clerical officers, these good fellows (our representative,
Mr. Smith, will call upon you . . .), herding together for
protection against their own individual nonentity, uneasily
cocky in their own century of the Common Man—what had
they to support them but a kind of abstract and meaningless
status? "Unreal City," he murmured: "A crowd flowed over
London Bridge, so many, I had not thought death had undone
so many."

Not that a publisher's office gave any strong conviction of
reality, either. How could it, living on such an unsolid commodity as words? Upon authors, those furtive and self-centered
ghosts forever spinning words to conceal their shame, their

27

impotence, their inescapable shadowiness? Miss Miles, who wrote far too much; Stephen Protheroe, who had apparently dried up; General Thoresby—no, the General was different, a man who used words, as he would use artillery, to destroy the enemy.

"Do you believe in coincidence?" Nigel suddenly asked a total stranger—one of the black-hatted brigade—who had sat down at his little table and was unspeakably pouring ketchup over a plateful of cold salmon.

"Pardon?" The individual eyed Nigel with suspicion and resentment.

"I said, do you believe in coincidence?"

"Worl, yew've got to be a bit cautious nahdays whatchew credit. And who yew talk to," he added truculently.

"Throw caution to the winds. It's the bane of you civil servants. Forget your files for once, and study a live human face. Do I look like a confidence man or a Teddy boy?"

"Worl—"

"You are perfectly correct. I can see you are an acute judge of character. Now then, do you believe in coincidence?"

"Funny thing you should mention it," replied the stranger, humoring him. "Only this morning the worf says—"

"Never mind what your wife said. I'm asking you—"

"Aw downt quart lawk yewer tone."

"In an age swamped by mechanistic physics and mechanistic psychology, the only rock left above the surface is coincidence —beautiful, anarchistic coincidence. In a society that bows down and worships at the altar of statistics, coincidence is the one remaining manifestation of a higher Providence."

"Aw'm a free thinker mawself."

"You will say, perhaps, that the science, art or sullen craft of criminal detection should confine itself to facts which admit of a causal explanation. I disagree. A science which leaves no room for coincidence—that is to say, for two apparently related events to happen simultaneously without there being any actual connection between them whatsoever—is an inadequate science, a false science. Must you go?"

The stranger mumbled that he had seen a friend come in, and removed his plate, his beer and his person, the latter visibly sweating.

Having got the table to himself again, Nigel resumed his train of thought. For all his recent defense of an arbitrary and overriding law manifesting itself in coincidence, he found it difficult to swallow the particular example he had been offered this morning. He turned once again to the offending passage in General Thoresby's book. There it was, crossed out, a delete mark at the side, this mark crossed out in its turn, dots under the lines and a stet sign clearly written in the margin. It was not a very long paragraph, but it was dynamite.

To sum up, the Governor's handling of the disturbances from start to finish (if "handling" is a word appropriate to one who never lifted a finger till it was too late) gave an object lesson in nerveless ineptitude, and constitutes a major blot upon the already checkered annals of our Colonial Administration. Firm action at the start would have rendered the disaffected elements powerless. Firm action, even after the rising had gathered force, would have quelled it with a minimum of bloodshed. But the Governor, occupied with the more congenial business of cocktail parties, opening bazaars and enjoying his siestas, took no action whatsoever except to obstruct the military in their efforts to disperse the mobs. As a result of his criminal negligence there was considerable loss of life and widespread destruction of property. The holocaust at the Ulombo barracks, where a half company of the Sussex Fusiliers perished to a man, is but one of the needless disasters for which a supine and incompetent administrator must be held directly responsible.

Whether or no they were fair criticism, Nigel's heart warmed to the author of these cannonading periods. The chapter which they concluded began with an account of General Thoresby's period of service as O.C. the troops in the colony. He had clearly been at loggerheads with the Governor-

General over the steps which should be taken in view of its unsettled condition, and had been recalled to England just before the disturbances actually broke out.

. . . And Millicent Miles, before she had realized that Stephen Protheroe was not alone in his room, had asked him how to spell "holocaust." . . .

What Nigel had so far read of *Time to Fight* compelled respect both for the General's professional abilities and his literary style. When he used the word "holocaust," he would use it in its correct modern meaning of "complete destruction by fire"—perhaps, considering his indignation against Blair-Chatterley, in its original meaning of "a sacrifice wholly consumed by fire."

General Thoresby's book showed him a brilliant soldier fanatical over his own conceptions of strategy, tactics and logistics. Nigel turned back to the first of the two libelous passages. Here, describing a certain push in France in 1944, the General had written,

The offensive was entirely successful, except on the Paume-Luzières sector. Here, during a mercifully brief period of command, General Blair-Chatterley's Crimean and costive conduct of operations achieved 2,586 casualties among his own troops and a noticeable heightening of the enemy's morale.

All but the first sentence of this passage had been deleted, then stetted. Though one might fault it for excessive alliteration, its panache made Nigel all the more eager to meet its author. An appointment at the General's club had been fixed by telephone, and thither Nigel now betook himself.

A waiter showed Nigel to the library, where General Thoresby was sitting alone, deep in a volume of Proust.

"Ah, Strangeways? You drink, I hope? Two Armagnacs—large ones."

A small, erect, dapper man shook hands with Nigel. The General's face presented an odd combination of the don and the pirate: gray hair, a high forehead, dreamy blue eyes; but

beneath them the swashbuckling mustache, full red lips, a
ram of a jaw. His voice was gentle, but turned staccato at
moments of excitement.

Nigel complimented him, with sincerity, on what he had
read of his book.

"Surprised to find a soldier literate?" The General chuckled.

"I'm surprised to find anyone literate nowadays."

"Geraldine tells me you're looking for the culprit. Don't
know if I encourage that. Good luck to the fellow, I'd say."

"But you didn't creep back into the office and do the dirty
work yourself, sir?"

Thoresby chuckled again. "I'd have done it if I could, but
they'd have spotted me. Of course, I might have bribed one
of Geraldine's minions to do it for me. Had you thought of
that?" The blue eyes gave Nigel a schoolboy's mock-innocent
stare.

"As a matter of fact, I had."

"But I didn't. Word of honor. Ah, here are the drinks."

General Thoresby lifted his glass to Nigel. "Good health.
I won't say good hunting. I see you're trying to sum me up.
Don't be misled by my mustache. Grew it to impress the
troops, and can't be bothered to remove it now. Camouflage.
Goes well with a bowler hat, too."

Nigel took the reference. "You were retired after your row
with the Governor-General?"

"Yes."

"Do you mind if I ask you a very straight question, sir?"

"Ask away and I'll see."

"Your attacks on General Blair-Chatterley in this book—
were they, well, you know, pro-bono-publico criticisms, or
have you some personal feeling against him?"

"A bit of both. He needs showing up. Also, one of the two
young officers burnt alive with their men in the Ulombo barracks was a protégé of mine."

There was a pause. The General frowned fiercely at the
volume of Proust on the table beside him.

"I see," said Nigel presently. "That makes things more awkward still."

"How do you mean?"

"In the libel suit. The plaintiff will be able to allege a personal grudge on your part. Difficult for you to plead fair comment."

"The plaintiff might find that allegation cutting both ways," said the General mildly.

"He had a grudge against you? Before the appearance of the book?"

"My report on the political unrest in the colony, and the military dispositions that should have been made to counter it—not at all gratifying to old B.C."

"But he was—"

"Oh yes, he was whitewashed, after the damage had been done. A deal of influence in high quarters, has B.C." General Thoresby's voice became staccato. "A nice lot of whitewash. A nice bit of hushing-up. Reasons of high policy. Politicians make me vomit."

"I see."

"What do you see, my boy?" The blue eyes were piercing now, and alight with intelligence.

"You *want* this libel case to go forward. You want Blair-Chatterley's bungling brought out in the open—all the whitewash scraped off—and a libel action is the only way it can be done. Even though it could ruin you financially. I call it rather public spirited."

"I can assure you of one thing. I'd have done just the same if that young friend of mine had not been killed out there."

"Yes. I'm not trying to be censorious; but Wenham & Geraldine, and the printers—they never asked to be dragged into this battle of yours."

The reckless, piratical aspect of General Thoresby now came uppermost. "They can take their chance. Publishers insure against libel, don't they? And they've got some sort of defense—they did their best to have these bits cut out of the book."

"That's no defense. They want to settle out of court, you know."

"Old B.C. won't consent," replied the General with glee. "He daren't. The thing's spread around too far already. He's got to come and play in the mud. Have some more brandy."

"No thanks. And your defense—?"

"I've had it out with my needle nose. He's instructed to plead that I told the truth in those passages, and told it for the public benefit."

"Well, it's your funeral. You can support your statements?"

"My dear chap, I've been occupying much of my enforced leisure during the last few years in accumulating evidence to support my charges. I've got quite a dossier on old B.C. Even my needle nose was impressed. I may be mad, but I'm not a fool."

Nigel lit a cigarette. "Good luck to you then, sir. But I'm afraid all this gets me no further with my little problem."

"Let's have the picture."

Nigel gave it to him, briefly. The General's quick intelligence appeared in his comment: "So almost everyone in the firm had the opportunity. You'll have to go for motive then. Some vindictive type, bless his heart, who wanted to do me down and didn't realize he was doing me a favor."

"Or wanted to injure the firm. Yes." Nigel gazed round at the fusty-looking books which lined the walls. "Did you ever come across a woman called Miles—Millicent Miles?"

"What? The writer? Did I *not*? Absurd woman. Smacked her bottom for her once—metaphorically."

Nigel asked for more. General Thoresby went on,

"Early in 1940, it must have been. My battalion was sitting about on the east coast, waiting for the real war to begin. Troops getting browned off. They sent us some lecturers, including your Miss Miles. She came down and gave a talk about novels. Literature-for-the-tots stuff. Condescending. She bent down, so far to our pathetic intellectual level you could hear her stays creak. And she dragged in some palaver about how wicked war is. Bad form, under the circumstances. Nobody knows better than the professional soldier that war is about the bloody silliest pastime ever thought up by humanity. And the troops don't like you patronizing them and they were champing with fury before she was halfway through." A reminiscent gleam lit up the General's eye. "So afterward, when she was dining with us in the mess, we did rather take her to bits."

"I'd like to have been there."

"The lady's not a friend of yours? Good. Well, we had some clever chaps; and I read myself, y'know. So we launched a counter-demonstration, a real highbrow free-for-all. The air was thick with chunks of Henry James, Proust, Dostoievsky, *Finnegans Wake*—the whole works. The Miles woman simply couldn't compete. Hide like an ox, of course; but after a bit she realized what was happening to her. Didn't like it at all. Outrage by the brutal and licentious soldiery. Refined female exposed in all her intellectual nakedness. Handsome sort of woman though, if you like horses."

"Have you met her since?"

"No, thank God."

"She still calls you 'Thor.'"

"Still calls me—? Well, the infernal impudence! Never seen the woman, before that occasion or after. Why are we talking about her, anyway?"

"She was working at Wenham & Geraldine's when your proof was interfered with."

This seemed a good curtain line, so Nigel rose to go.

"Well, good-by, my boy. Enjoyed our chat. Keep me in the picture. If you find the miscreant, I'll slip him a thank-offering; or her, as the case may be."

Nigel's next interview was of a very different nature. He had telephoned Mr. Bates, the late Production Manager at Wenham & Geraldine, saying he would like to discuss a business matter with him. Nigel did not usually care for conducting interviews under false pretenses; but in this case he judged it necessary.

Herbert Bates lived in a villa at Golders Green. His black suit, high stiff collar, and rather lugubrious face gave him the appearance of a family retainer—which, in a sense, he had been. His manner, too, was ceremonious, discreet, respectful. A hushed personality.

"If you will come into the lounge, Mr. Strangeways. I think you will find it warm and comfortable in this inclement weather."

Mr. Strangeways explained that, having recently received a large legacy and always been interested in publishing, he was thinking of setting up for himself in the business, and in the meantime had joined Mr. Bates' old firm to get practical experience.

"And no better house could you find for that purpose, if I may say so, Mr. Strangeways."

"It's all very much in the air at present. But, if my plans go forward, Mr. Bates, would you consider taking up the post of Production Manager for me? Miss Wenham speaks of you in the highest terms."

Nigel was extremely relieved to notice that no gleam of eagerness appeared in Mr. Bates' eyes.

"I greatly appreciate the suggestion, sir, and Miss Wenham's kind recommendation," Mr. Bates replied in his tired, toneless voice. "But I fear that, at my age—"

"Oh, come now, Mr. Bates, you can't be sixty. Surely you don't want to retire permanently yet?"

"Sixty-one next birthday. I had intended to stay on with the firm till I was sixty-five; but—er—events transpired which made a somewhat earlier retirement—ah—feasible."

Nigel nodded gravely at these diplomatic circumlocutions. "I am sure Miss Wenham and Mr. Geraldine must have greatly regretted your decision. Mr. Ryle's acquaintance I have hardly made as yet; my impression, though, is that he does not quite—" Nigel left the sentence dangling delicately in the air, and Mr. Bates came to the lure after a moment's hesitation.

"No doubt he will learn Our Ways in good time, sir. I confess I did not always see eye to eye with him myself. A publishing house is not, after all, a factory. Ah well, youth will be served, as they say."

Mr. Bates gave a resigned sigh. If this man is capable of bearing resentment, thought Nigel, then thistles can bear figs. Further gentle probing confirmed him in this impression, so he went off on another tack.

"I've just been having a talk with General Thoresby," he said. "Interesting man."

Mr. Bates looked severe. "Disgraceful! I should have expected better from a man of his standing. But you can never trust authors—slippery customers, the lot of them—and no gratitude either. A deliberate attempt to involve us in a libel action."

"The attempt has succeeded."

Mr. Bates was flabbergasted, not having heard of the most recent developments, which Nigel now summarized for him.

The expression on Mr. Bates' face was that of an old family retainer who has found a skeleton in the family cupboard—shocked, but a trifle prurient.

"And the partners think it must have been done by someone in the firm, you say, sir? God bless my soul! The imagination boggles at such an enormity. It is quite unprecedented in Our Annals. I well recollect Mr. Protheroe bringing in the proof copy that morning. After all that trouble with the author, and falling behind schedule, I was glad to get it off to press. I remember passing a remark to that effect, while my secretary was typing a note for the printer and parceling up the proof; and Mr. Protheroe—he was sitting beside me, as it might be the way you are now, sir, except of course that I was at my desk—Mr. Protheroe responding with a witticism to the effect that the soldier's pole had at last fallen—an expression first used, of course, by the Swan of Avon."

"How very interesting"— Nigel's comment referred to the matter rather than the manner of this discourse. "So Mr. Protheroe was with you all the time you were having the proof dispatched?"

"Just so, sir. A matter of a few minutes perhaps. But we had a very pleasant little chat, I recollect. Conversation with Mr. Protheroe I found all the more agreeable for its being a rare occurrence. An able mind there, Mr. Strangeways: whimsical; a little eccentric, perhaps: but keen. Oh, a fine judge of literature is Mr. Protheroe. It will be a sad day for Wenham & Geraldine when Mr. Protheroe reads his last manuscript."

Nigel thought, rather irritably, that all this trouble could have been saved if Stephen had told him he was present when *Time to Fight* was dispatched from Mr. Bates' department. The late Production Manager could clearly be eliminated from suspicion. He could not so easily be shaken off, however. Combining the upper servant's passion for gossip with the power to make it sound like the stately exchanges of old-fashioned diplomacy, Mr. Bates offered no conversational loopholes through which the visitor could escape.

If he could not be stopped, though, he could be switched. And Nigel switched him onto the present partners in Wenham & Geraldine. Liz Wenham, he already knew, was the founder's granddaughter. Arthur Geraldine was a great-nephew of the original John Geraldine. Though Arthur was technically the senior partner, Liz Wenham in Mr. Bates' view provided the firm's mainspring: she had her grandfather's head for business. She was not always, however, quite so happy in her dealings with authors (Mr. Bates' lips primmed at the mention of these distasteful but necessary adjuncts to a publishing business). Here was the strength of Mr. Geraldine, who had "a way with him" and conducted much of the personal negotiations with the *genus irritabile*. Mr. Geraldine also possessed an almost uncanny flair for estimating in advance the sale of a book. This, Mr. Bates gratuitously pointed out, is the most essential requirement for success as a publisher: print too large a first impression, and you are left with a load of unsold stock on your hands; print too few at the start, and you may be unable to reprint soon enough to satisfy the un-expected demand. Either way, money is lost. Mr. Geraldine had the priceless faculty for preassessing the public response in terms of copies sold.

After a number of anecdotes illustrating this gift of Mr. Geraldine's, Mr. Bates moved on to the new partner. Basil Ryle had replaced a Mr. Charles Wainwright, who was killed in a motor accident the previous year. Ryle had started his own publishing business after the war. It had promised well at first; but rapidly rising costs, and the lack of a good solid backlist, had defeated him. Last year, when it became evident that he was nearing the rocks, Wenham & Geraldine had be-gun negotiations with him, finally taking over the stock and good will of his business and giving him a partnership.

"Was Miss Miles part of his stock, so to speak?" asked Nigel.

"He had not published her; but I believe he had com-missioned her autobiography—or at any rate, was negotiating for it."

"A valuable property."

"In terms of potential sales, I agree."

"But not in any other terms?"

"The author's life has been, I am told, somewhat irregular," pronounced Mr. Bates.

"Her book," Nigel could not resist asking, "might be cal-culated to bring a blush to the young girl's cheek?"

"I doubt if it will prove to be wholesome family reading," Mr. Bates assented. "You know, perhaps, that she is a divorcee?"

Nigel expressed suitable horror.

"Married three times. Not to mention—hr'rm."

"You don't say so! Any children?"

"There is a son by the first marriage. Cyprian Gleed. A ne'er-do-weel, I regret to say. Gave us great trouble a few months ago."

"What? Gave the firm trouble, you mean?"

"Yes. He wanted us to back him in starting a literary periodical. He's one of those young fellows who never settle down to anything—Bohemian—a hanger-on of the world of letters."

"He was refused?"

"Certainly. Miss Wenham and Mr. Geraldine were rightly adamant. But he pestered them for quite a time—kept turning up, with or without an appointment. His mother made it all the more difficult: she had been lent a room to work in, and we could not very well refuse him access to *her*."

"When did this happen?"

"A few months ago. Let me see, last July it would have been."

"You don't remember if he paid one of those visits the day *Time to Fight* was sent to press?"

"I'm afraid not. But, if you are interested, Miss Sanders could tell you—our receptionist. She makes a note of every visitor."

IV *VERSO*

Nigel got back to Angel Street just before the main door of Wenham & Geraldine was shut. He had glanced through the evening papers on the Tube: one carried a review of *Time to Fight*, mentioning but not quoting its strong criticisms of Sir Charles Blair-Chatterley; another had a news story about the injunction and the withdrawal of the book, followed by a brief account of the trouble in the Colony after General Thoresby had been superseded as O.C. the troops there. The third made no mention at all of the book or the impending libel action.

The gipsyish Somerville graduate was at her desk in the reception room. Nigel, who had decided that he could not ask her for the information he wanted in his role of specialized reader, said straight out, "Mr. Geraldine tells me you keep a record of all visitors to the office, Miss Sanders."

"That is so."

"Would you please tell me who came in during the afternoon of July 23rd and the morning of July 24th."

The girl looked mystified, and a little distressed. "I'm not sure if—I mean—"

"Ring Mr. Geraldine and ask him if it's all right for you to give me the information."

She did so, then took out from a drawer a large leather-bound book and turned over the pages—rather slowly, as if she were also turning over some problem in her mind. "July the 23rd, you said? Here we are. Afternoon. General Thoresby,

Mr. Leeson-Williams, Miss Miles. . . ." She recited a list of some dozen names. "July 24th. Mr. Ainsley, Miss Miles, Mr. Bellison. Mr. Smith, Mr. Ritchie, Mrs. Vane, Miss Holloway, Rev. Dowle. That's up to midday on the 24th."

Nigel had noticed an almost imperceptible pause between the names of Bellison and Smith, but made no direct comment on it. "That's all, is it?" he asked.

"That's all." Her eyes met his with a bold, level look, as she dropped the book back into the drawer. If she had any curiosity about Nigel's odd request, she refrained from showing it.

Nigel thanked her, went out of the room, rang for the lift, took it up to the first floor, quietly got out and came down the stairs.

"Left my evening papers here," he said as he entered the reception room again. Miss Sanders, taken by surprise, had no time to conceal the leather-bound book or the ink eraser. Coloring furiously, she resisted for a moment his attempt to take the book from her; then, with a shrug, yielded it.

Not looking at the book, Nigel asked, "Why were you rubbing out Cyprian Gleed's name?"

"I wasn't doing anything of—"

"That's not very bright, for a First in History." Nigel opened the book at July 24th, exhibiting, between the entries of "Mr. Bellison" and "Mr. Smith," the half-erased name of Miss Miles' son.

"Who the hell are you?" the girl angrily exclaimed. "An efficiency expert or something? It's absolutely intolerable—"

"Why were you rubbing out Cyprian Gleed's name?"

"I refuse to tell you."

"Just don't like the name? I'm not surprised."

This provocation had the desired effect. The girl blurted out, "Cyprian is— If you must know, he came to see me privately. The firm doesn't encourage followers during office hours, and I didn't want you to go tattling to the partners about it, as you seem to be some sort of spy."

"If it was a private visit, and such visits are frowned upon, it's strange you should have entered his name in the book at all."

Miss Sanders flushed again. "It becomes quite mechanical," she said.

"Even with 'followers'?"

"You're foully offensive!"

"It was your word, not mine. How long was Mr. Gleed in the building, do you remember?"

"He was in this room the whole time," the girl protested.

Nigel made no comment on the wording of her reply.

"Do you remember the exact time? . . . No, for heaven's sake don't lie about it. Presumably Mr. Smith and Mr. Bellison had appointments, and that would fix the time Cyprian Gleed arrived."

"He came in around half past ten, I think, and stayed for—no, I can't remember—it wouldn't have been very long. And now, what *is* all this about?"

"My dear girl, you know quite well what we're talking about."

"Cyprian is a man of absolute integrity."

"Good. Then why need you get into such a state? A man of integrity would never have played that trick with *Time to Fight*"—Nigel held up the proof copy—"to get his own back on the firm for their refusal to back his literary magazine. Would he?"

"Everyone misjudges Cyprian," said the girl miserably. "If they'd only give him a chance—"

She took off her spectacles and began polishing them vigorously: her eyes looked bleared and strangely naked.

"Well," said Nigel, smiling as he handed her back the leather-bound book, "I must be about my sinister business. Cheer up. All is not lost. Not yet, anyway."

Nigel plodded upstairs to Stephen Protheroe's room. Stephen gave him a preoccupied glance, then buried his head again in the typescript he was reading. He appeared to read with his nose, which hovered over the pages like a humming-bird hawkmoth, darting down now and then as if to smell the author's style, twitching and sniffing frequently. After sampling thus several pages of the book, he exclaimed "Faugh!" scribbled a few words on a sheet of paper, attached it to the typescript, and looked up at Nigel again.

"They say that every man has one good book in him. They are wrong. Where've you been?"

"Talking to people. Mr. Bates, for instance."

"And what was he buzzing in your ears?"

"He said you were present when he sent *Time to Fight* to press."

"If he says so, no doubt I was. Why?"

"Well, he couldn't have tampered with the proof: you'd have seen him."

"But I told you he wouldn't have done it. Bates is a bore, but not otherwise flagitious."

"There's a difference between 'wouldn't' and 'couldn't.' If, as he claims, you took the proof into his room and stayed there talking while a covering note was written and the proof wrapped up, your evidence clears Bates."

"Good. Then I'll give that evidence."

"But do you actually *remember* that it was so?" asked Nigel patiently.

"Too long ago. But I'm sure Bates is right. He is totally devoid of imagination and therefore he could not have imagined it."

With this Nigel had to be content. Stephen's eye was already straying toward another typescript.

"Did Miss Miles' son visit her that morning?" Nigel asked.

"That young clot? May have, for all I know. He was infesting the office last summer."

"You don't happen to have a note of the date his proposition was turned down?"

"No. I'll ask." Stephen dialed a number on the internal telephone. "Arthur? Stephen. What date did you and Liz turn down young Gleed's horrible scheme for a literary mag? You had a final meeting with him, didn't you? . . . July 21st. Thanks." Stephen turned back to Nigel. "July 21st. Sinister coincidence, eh? Oh, I forgot. Ryle said he'd like a word with you when you got back."

Nigel left the little man, his gnomelike head floating in a pool of yellowish light from the unshaded bulb, and went down to Basil Ryle's room. The junior partner was at his desk, conferring with a depressed-looking middle-aged woman.

"Just one moment, Strangeways. Take a chair." Ryle handed the woman an advertisement pull. "This was yours, Miss Griffin? Look at it again. Do you really think any living soul is going to be attracted to the book by this advertisement? 'A charming novel about young love, in an East Anglian setting.' Sets the blood on fire, doesn't it? Fills one with seething curiosity."

"Well, I'm sure I—"

"We pay £25 for the space. If it doesn't catch the reader's attention, we might as well give the money to a home for lost dogs."

"What about 'a broad on the Broads'?" suggested Nigel.

Miss Griffin received the suggestion unsmilingly. "I don't think Mr. Geraldine would care for—"

"We're not selling the book to Mr. Geraldine," Ryle broke in. "We're trying to sell it to a lot of dopes, who'd much rather be looking at television. Wait a minute—what's the heroine's name? Helen. It *would* be. 'Nelly or Telly'? Something on those lines? No. They'd choose Telly every time. I've got it—'The tale of a Suffolk Helen who broke all hearts.' "

"But actually, Mr. Ryle—"

"I daresay she didn't. But ninety per cent of novel readers are women, and ninety-nine per cent of women would like to think they leave a trail of broken hearts behind them. All right, Miss Griffin."

Miss Griffin retired, with a look that suggested the breaking of heads rather than of hearts.

"The trouble with this firm is advertising in a sort of well-bred undertone, like a butler offering red wine or white." Basil Ryle ruffled his mop of ginger hair, and suddenly looked very young indeed. "Well, how've you been getting on?"

Nigel gave him a censored account of his interview with General Thoresby and a fuller one of the talk with Mr. Bates.

"You can count Bates out, I think."

"Yes, I suppose so." Ryle bit his nails. He seemed jumpy: no doubt, being a live wire he would. "But what about clues? Fingerprints, I mean? Handwriting? I'd have thought that's the first thing you'd be after."

"I've asked one of the handwriting experts at Scotland Yard to look at the proof this evening. He's a friend of mine. But one word, written twice in block capitals, isn't likely to get us anywhere. Fingerprints? The proof copy will be covered with them, for one thing. Besides, do you think the partners will let me fingerprint everyone in the building?"

"Why not?"

"Including Miss Miles?"

"Millicent? How does she come into it?"

Ryle's tone was sharp. Under stress, a slight coarseness could be heard in his vowels. Provincial boy, from lower-middle-class parents, who had made good—that accounts for the way he throws his weight about with subordinates like Miss Griffin, thought Nigel; he's touchy, and will need careful handling. He studied for a moment the red-haired young man, who had tilted his chair back against the wall—the neat, dark suit, the soft collar and tie rumpled.

"Miss Miles had a motive for doing it; two motives, actually."

"Motive? Come off it."

"She had been humiliated in public, during the war, by General Thoresby. And this firm has refused to reprint some of her novels."

"You call that a motive?"

"For an egotistic and vindictive woman it could be."

Basil Ryle pushed himself from the wall, the front legs of his chair coming down with a bang. "You set up as a professional judge of character, do you?" he dangerously began. "Now see here— Yes, what is it?"

An attractive blonde had entered the room.

"Your letters, Mr. Ryle."

"Oh, damn the letters!"

"You said you wanted them to go off by the six o'clock post. It's five to six."

Basil Ryle started signing a sheaf of letters in a folder on his desk. "You want to meet the boy friend at six, I suppose."

This bit of badinage sounded curiously artificial and unconvincing, as though he had learned it from a correspondence course. The blond secretary was looking down at Ryle's head in a maternal, almost intimate way. He's all thumbs where girls are concerned, thought Nigel, yet they feel an attraction.

" 'Colophon' has only one 'l,' Daphne. All right, I've altered it. . . . Here you are. Run along, and don't let him take any liberties."

"Good night, Mr. Ryle."

When the girl had left, Basil Ryle fidgeted with the pulls on his desk, then jerked out, "You free for a bit? Let's go and have a drink. I'm sick of this office."

They left the building, silent now and with only one or two passage lights burning, by the side door. The night air was clammy cold, but Ryle inhaled it vigorously.

"That's more like it," he said; then, cocking his head toward the Thames, "Never thought I'd want to see a river again. My dad was a welder on Tyneside. One of the casualties of the Slump. Just rotted quietly away. Poor old dad." In the lamplight, as he glanced up at the elegant façade of Angel Street, his face wore a pleased, incredulous look, as though he could hardly believe yet in the fortune which had brought him here. "My dad was a great reader. Plenty of time for it, too, after 1930. And the public library was warm. But you can't eat books."

He led Nigel into the private bar of a small pub, tucked away between Angel Street and the Strand. "Not many know of this place. We can have a quiet talk." The bar was indeed empty; and with its high-backed settles and cheerful fire, it had a countrified feel. Ryle ordered a double whisky and chaser for himself, Hollands for Nigel.

"Millicent Miles," he said abruptly. "You know her well?"

"I only met her today."

"Well, I do. And I'm telling you, she's a greatly misjudged woman. She had a rough time when she was young. Like me. It scars you. It's all very well for a highbrow like Protheroe to turn up his nose at her books. Of course, they're escape stuff—"

"Escape from what?"

According to Basil Ryle, Millicent had been the daughter of a drunken father, who went bankrupt when she was in her early teens, and a slut of a mother. Her parents quarreled incessantly; the father's temper was unpredictable—he would beat the small girl, then drool over her in alcoholic remorse; the mother used her as an unpaid slavey. Millicent had run away from home at the age of seventeen and got a job as a shop assistant. Her miserable and repressed girlhood had generated the fantasies which later made her the darling of the lending libraries.

"You've been reading her autobiography?" asked Nigel.

"No. She told me all this." During his narration, Ryle had put down several whiskies, and the effect was beginning to tell. "You probably think she's a hard-boiled number. You're quite wrong. Underneath, she's a damned sight more vulnerable than most of the sensitive plants of both sexes you meet."

"When did you first meet her?"

"And lonely too. At one of those literary cocktail parties. Some years ago. Before I was in business on my own. I said to her, straight off, total stranger, 'You look lonely.' Extra-

ordinary—don't know what possessed me to do it. And she
said, "You're the first man who has realized that.' I could see
she'd had a rough passage with men. Don't believe everything
you're told about her."

"I won't."

"What's that?"

"I said 'I won't.'"

"Good for you. Have another drink. No, I insist. Hollands
again? . . . Where were we? Yes. A rough passage. Do you
know, when she was nineteen a man seduced her; promised to
marry her—the usual story—then left her in the muck; baby
was stillborn. Oh, she's been through the hoops. Married a
stockbroker or something a year later, just to get away from
poverty. That's when she started writing. To get away from
the stockbroker, if you ask me."

"She has a son by him, hasn't she?"

"Cyprian. Yes."

"What's he like?"

"A sponge." His drinking had carried Ryle rapidly through
the animated to the melancholy stage. "A sponge with a beard.
Sucks up her money like water."

"Was he very annoyed when Wenham & Geraldine turned
down his project for a magazine?"

"I expect so." Ryle was already uninterested in the egre-
gious Cyprian. "That's one reason why I wanted the firm to
reprint some of her novels. She needs the money. She's absurdly generous, you know: offered to put some into my own
business, but it was too late by then."

"But there are other reasons?" suggested Nigel.

Basil Ryle glared at him owlishly. "What a damned lot of
questions you ask. Like one of those Bloomsbury relics. Always
probing and prodding away, as if a personal relationship was a
piece of meat on a grill. You in the war?"

"Not to speak of."

"I saw a crew after their tank had burned up. Talk about
grilled! You know what's wrong with your generation? You

believe in goodness, kindness, decency. Other things being equal, decency will always prevail—that's what your lot think. It's bloody pathetic."

Nigel let that one go past. He got Ryle talking about his start in publishing. Immediately after the war, he had gone into an advertising firm and quickly worked his way up. Millicent Miles' third husband was one of their clients—a wealthy, literary-minded manufacturer; he took a fancy to Basil, discovered his ambition for publishing, and found him the necessary financial backing to start a publishing business.

"It was like a fairytale," said Ryle. "But it didn't have a fairytale ending."

Nigel could imagine the tough, provincial young man flattered by Millicent Miles' interest, dazzled a bit by the world into which she introduced him. And no doubt she had kept him dangling: she couldn't do without a string of admiring men. But how had she kept him so long? Concealed her essential hardness and egotism so long from him? He must have a soft streak in him, a romantic streak; or perhaps it had something to do with the decency he claimed to disbelieve in—a loyalty which kept his eyes shut.

Leading the conversation back to Millicent Miles, Nigel learned that she had divorced her first husband—the stockbroker—who was now dead; her second, a neurotic racing motorist, had shot himself; the third, Basil's patron, had divorced her a year ago.

"She feels she is doomed never to make a go of it, never to be happy for long," said Ryle. "Poor girl. She *is* a girl still, you know, underneath all that camouflage."

Oh dear oh dear, thought Nigel, so you are in love with her, you're to be Mr. Right and pick up the pieces and put her together again.

"That's another reason why I'd like to reprint a few of the novels," Ryle continued. "It might help to set up her morale. And, as I say, with that leech Cyprian around, she needs money."

"But the other partners are against it?"

"Geraldine doesn't object. Liz Wenham is pretty lukewarm, I admit. But Protheroe's the real obstacle."

"Surely he doesn't have all that say in what you publish?"

Getting up, Basil Ryle poked the fire viciously, then leaned on the mantelpiece, staring without enthusiasm at an advertisement of a naked young woman offering the public a bottle of sparkling cider.

"Protheroe's been the firm's literary adviser for twenty-five years," he said at last. "And it's true he's never turned down a book which some other firm then accepted and had a success with. I give him that. But he's getting a bit fusty now; won't look at anything that isn't what he calls 'written'—not in the fiction line. Well, fine writing's on its way out. Anyway, the answer is that the partners wouldn't take on anything he was dead against, unless they were both enthusiastic about it."

"They were enthusiastic about Miss Miles' autobiography, then?"

"Not wildly. But I'd given her a contract to write it while I still had my own business, so Wenham & Geraldine more or less had to take it over."

"Over Protheroe's dead body?"

"No. He didn't make a fuss about that, I believe—only about the novels. But he does seem to have a down on Millicent."

"Something out of the past? Perhaps they knew each other—"

"No. I asked Millicent, and she said she'd never met him till last summer. Dried-up little runt. What's *he* ever produced? One slim volume of verse."

"A great one, possibly."

"Do you think so? I've read it. All lust and disgust. Venus and vinegar." Basil Ryle slowly applied the lighted end of his cigarette to the belly of the girl on the advertisement. "Sex!" he muttered. "Why can't they keep it for the dark?"

V RUN ON

Next morning, Thursday, Nigel Strangeways entered Angel Street a little before half past nine. It was another of those gray days, the skyline a dirty dishcloth, houses and trees, the river and its bridges all looking as if they had been smeared over with a film of grease. A metallic cold in the air outside; and a nip in Miss Sanders' voice as she returned his "good morning." Mr. Gleed, she bleakly announced, wished to see him, and was coming at eleven o'clock for that purpose.

"Good. That'll save me a lot of trouble. Ask him to leave his knuckle-dusters with you for safekeeping."

Miss Sanders frowned at this flippancy. One could imagine the disapproving look she might have cast upon a don who had let slip some frivolous remark on a serious subject during a tutorial. How very *stern* the young are, thought Nigel as he went upstairs; one keeps forgetting it. Basil Ryle last night with his censorious line about sex: but that's the old-fashioned working-class puritanism; and anyway he can't be much over thirty; but emotionally undeveloped?—time will show.

Stephen Protheroe was already at work, nosing his way through a bulky typescript, and returned Nigel's salutation absently. A word caught his attention. His long nose dipped at the page like a woodpecker's beak.

"Another fellow who says 'disinterested' when he means 'uninterested,'" he snarled. "This debasing of the language is intolerable."

"Words have changed their meanings in the past."

"That's not the point. The point is, there's no synonym for 'disinterested.' We can't afford to lose the word."

"Perhaps we've lost what the word means."

"Lost disinterestedness? As an ideal? I wonder." The little man's face was suddenly sad—ravaged by a sadness Nigel had rarely seen: the blank eyes, the downturned corners of the mouth changed it into a tragic mask. It was scarred with deep furrows—riverbeds grooved out by emotional torrents long ago run dry.

"You haven't got a *Who's Who*, have you?" Nigel asked.

Stephen dragged himself out of whatever pit of sorrow, regret, remorse he had been plunged in. "The last room down the passage has a reference library," he said.

Nigel went out, passed Miss Miles' door (the typewriter within was silent), and entered the large room beyond it. A young, disheveled but attractive blonde, considerably less scholarly-looking than most of the firm's female employees, was combing her hair at a pocket mirror.

"Oh," she uttered, in a sort of refined yelp. "You quite took me by surprise."

"Please don't get up. My name's Strangeways, and I'm working in Mr. Protheroe's room. I came to look for a *Who's Who*."

The girl appeared to be dazed by this request.

"Well, I don't know if—you see, I'm not really here."

"You mean, you're a ghost?"

The girl's blue eyes opened incredibly wide. "Ooh no, I assure you. I'm reelly downstairs."

"You amaze me."

"I work in the invoices. But Jean's sick, so I'm here instead. What was it you would be wanting?"

"*Who's Who*."

"Pardon? Whose Zoo?"

"It's a reference book, called *Who's Who*."

"Oh, a *book*." The girl's eyes went desperately round the room, which was lined with books from floor to ceiling. "We

don't have anything to do with the books here." She sought
for inspiration in the ceiling. "Now Mr. Protheroe—he knows
about the books. Why don't you try him?"

"But Mr. Protheroe sent me here."

"Well, isn't that funny?"

"Perhaps if I looked round the shelves—"

"You're welcome, I'm sure."

"It's got a fat red back."

The girl giggled. "You are naughty."

Nigel glanced over the shelves. File copies of the Wenham
& Geraldine publications, going back over a hundred years. In
the far corner, by the window, two shelves of reference books.
Taking out the one he wanted, Nigel turned to the name of
Millicent Miles, and began jotting down notes. The blonde
leaned over his shoulder, enthralled.

"Would Johnny Ray be there?"

"I don't know. Who's he?"

The blue eyes opened so wide that Nigel felt in imminent
danger of being engulfed. "Who's Johnny Ray? He's— Go on!
You're sending me up!"

"Go on yourself."

"He's a singer. He's my heart throb."

"Lucky chap!"

The blonde, temporarily unfaithful to her heart throb,
pressed more heavily against Nigel's shoulder, as he turned to
another name in Who's Who.

"You know, you remind me of a film star—in one of those
corny old films—my boy friend took me to it—he's a highbrow,
my boy friend I mean. Now, what was his name? Burgess
Meredith."

"Don't tell anyone," whispered Nigel into her shell-like ear,
"but I am Burgess Meredith. Who are you?"

"Don't be silly! Oh, I see. Susan—Susan Jones. But honestly,
you're not reelly—?"

"Ssh!" hissed Nigel. "I'm actually a famous detective, a

private eye disguised as Nigel Strangeways disguised as Burgess Meredith."

The girl laughed merrily, shaking her white-gold hair against his cheek. "A detective! That's one thing you aren't—anybody could see."

When he had finished taking notes and replaced the volume, Nigel said, "You must be lonely here all by yourself all day. Still, you've got plenty to read."

"I *hate* books. You can have the lot, for me. Now my boy friend—he's a great reader: always got his head stuck in a book. It fair gives me the creeps sometimes."

"Waste of time, when he's got a girl like you."

"That's what I tell him. We're only young once, aren't we? Meaning nothing personal, I'm sure, Mr. Mer— Mr. Strangeways." She gave him a languishing glance. "I like the mature type, myself."

"You be careful, young Susan. I'm at the dangerous age. And all these books around us rouse the passions."

"Ooh, you are *rude*!" cried the delighted girl.

After a little more of this sort of thing, Nigel inquired if she had come into contact with Miss Miles at all.

"That stuck-up cow? Give me air! Ooh, she's not a friend of yours, is she?"

"No. Doesn't anyone here like her?"

"Between you and I, they say Mr. Ryle's a bit—you know. Shocking, I call it. Old enough to be his mother."

"And what about Mr. Protheroe?"

"Oh, *he* can't stick her. Everyone knows that. Soon after she started camping here, they had words."

"*Did* they? What sort of words?"

"I didn't hear it myself. But my friend Jean, she happened to go into Mr. Protheroe's room for a moment, and they were at it hammer and tongs next door. She was being ever so offensive to Mr. Protheroe."

"How?"

"Well, Jeanie hadn't time to catch on what it was all about.

But she heard Miss Miles calling him Goggles. Ever so cold and nasty: like Bette Davis. Well, I mean, he's no Marlon Brando, but—"

"Was Jean sure she was calling him that?"

"She heard Miss Miles say 'Goggles.' Twice. Extra loud. She didn't catch the rest. Except they were both steamed up. But, with those gig-lamp spectacles Mr. Protheroe wears—well, it stands to reason she was passing a personal remark, doesn't it?"

"It's not his nickname here?"

"Oh no. Some call him the Prawn, I believe."

"Goggles. Who wears goggles?"

"Well, motorcyclists. And—"

"You'll be all agoggled to hear, Susan, that Miss Miles' second husband was a racing motorist."

On this satisfactory exit line, Nigel departed, leaving Susan Jones with her eyes as big and blue as a Hollywood star's swimming pool.

Nigel spent the next quarter of an hour in a conducted tour of the building. He wanted to fix its layout clearly in his mind, and Stephen Protheroe professed himself happy to get away from his manuscripts for a while. Starting on the third floor—the editorial department, which included the reference library and a studio—they descended to the second, occupied by the partners, their secretaries, and the advertising department. The first floor housed the strictly business side of the firm, the Production Manager's office, the accounts department, the invoice room, and so on. Finally, on the ground floor, there was the reception room, the trade department, the very large room where, under neon lighting and to the strains of music from the Light Programme, the packers worked, and an even larger storeroom beyond it.

"So there it is," said Stephen when they had returned. "Simple but symbolic. On the third floor Enlightenment, on the first floor Mammon; and the partners sandwiched between the two."

Nigel had noticed that his guide was received everywhere with the deference which might have been given only to a partner; but also, that there was no kowtowing in this deference—Stephen was evidently popular with the staff; and as far as Nigel could tell, the staff seemed on good terms among themselves. Stephen had affected a certain vagueness as to what went on in other departments, but answered Nigel's specific questions without difficulty.

For the next half hour Nigel examined the folder on his table. It contained the data he had asked for yesterday: a list of the staff, their periods of service and their duties; and a sheaf of pension forms, on which each employee had written his name in block capitals, and his signature beneath it. Nigel had never expected to get a lead from this, and he did not. He had asked for the data mainly to give an impression of efficiency. He put the pension forms in a large envelope: his friend at Scotland Yard might be able to make something of them, though to Nigel one man's block capitals were indistinguishable from another's.

"Can graphologists identify block capitals?" asked Stephen Protheroe, who had looked up from his work at this moment.

"Almost impossible, I should think, on a single short word like 'Stet.' Of course, if you've got some eccentricity, like crossing your T's from right to left, or forming a letter with an upstroke which is normally done with a downstroke—they'd be onto that."

"I thought so. How can you set about an investigation like this, then?"

"Look for motive. And opportunity. That narrows it down a bit. Which reminds me—Cyprian Gleed is coming to see me shortly. Where can I—?"

Protheroe jerked a thumb over his shoulder. "In there, if you like. The Medusa seems to be exercising her charms elsewhere this morning. What does *he* want?"

"We shall see. Did you ever come across her second husband?"

"Young Gleed's her son by her first."

"I know. I mean the racing motorist."

"No. I'm not a habitué of the pits."

"Shot himself, I'm told."

"I wouldn't be surprised." Stephen Protheroe's mouth made its nibbling, fishlike movement. "Some commit suicide, others have suicide thrust upon them."

The telephone rang. Mr. Gleed was here to see Mr. Strangeways. "Send him to the room where Miss Miles works," said Protheroe.

The figure that entered was far from prepossessing: a young man, not more than five feet six inches tall, in stovepipe trousers and a stained duffel coat, his pasty complexion rendered still whiter by a straggle of black beard and a black sombrero hat.

"You're Strangeways, are you?" he aggressively remarked, teeth flashing white behind the beard. "What the devil do you mean by insulting Miss Sanders?"

"Sit down and I'll tell you. Have a cigarette."

Cyprian Gleed's hand went out automatically to the proffered case, then withdrew.

"Thank you, no. I don't like soft soap."

Nigel reflected that, judging from his appearance, Gleed did not like soap of any kind. The young man had flung himself into the armchair; he was trembling violently, and furious with himself for trembling, and therefore trembling all the more.

"By what right did you bully Miriam—Miss Sanders—into telling you—"

"No bullying. She told me a lie and I exposed it. I've every right to do so. The firm has called me in to investigate the recent trouble here."

"What trouble?"

"Surely your mother told you about the libel case—"

"Oh, *that*. So you think your Gestapo methods are justified by— But how am I supposed to come into it?"

Cyprian Gleed was the type of moral weakling, Nigel judged, who must always be whipping himself up into aggressiveness to conceal from others his own lack of self-assurance.

"You come into it," Nigel equably replied, "because you were in this building at a time when the proof copy could have been tampered with, and because a project of yours had been turned down by the firm a couple of days before."

Cyprian's voice was contemptuous. "I suppose a private detective, or whatever you call yourself, is bound to have the keyhole mentality. Do you really think I care a damn if some tinpot general gets himself into trouble?"

"Why did you come back here, *after* your project had been turned down?"

"I came to see Miriam."

"But then you went upstairs."

Nigel made it sound like a statement, though he had no proof of it as yet.

"What if I did? I'm allowed to see my own mother, aren't I?" Annoyed with his weakness in replying, Cyprian gave another spurt of anger. "And why the bloody hell should I answer your questions?"

"You're not compelled to. Why, for that matter, *shouldn't* you co-operate? Or have you got a vindictive feeling against Wenham & Geraldine?"

The young man's eyes rolled, a trick recalling his mother.

"If you really think a writer has nothing better to do than go about feeling vindictive over his rejection slips—"

"I'm sure I should."

Gleed was taken aback, but then responded to Nigel's sincerity, launching into a tirade about what he called "the Establishment." This, it seemed, was a formidable, reactionary, cunning and cryptic body of people—editors, publishers and writers—who operated like a Black Hand in the world of literature, promoting their own interests and blocking the efforts of anyone outside the circle. It was men like Stephen

Protheroe, he declared, who were the kingpins of this racket—
éminences grises controlling literary policy, directing trends,
giving the jobs to the boys, assassinating the outsiders. The
Establishment was based in London, but had its agents in the
provinces too; it controlled not only publishing the literary
pages of the metropolitan newspapers and weeklies, but also
the B.B.C., the Arts Council, the British Council and the
older univeristies. It was all the more of a menace because its
members were, or professed themselves to be, unaware of its
existence. A self-elected oligarchy, they subtly extended their
influence or ruthlessly protected their own interests, but with
no greater individual consciousness of their corporate aims
than is possessed by coral insects or particles of smog. It was
to expose and combat this sinister, if amorphous, body that
Cyprian Gleed had proposed to start a literary magazine.

He spoke with genuine conviction; but Nigel found his
views none the less tedious, and finally broke in,

"Well, that may be so. But it's outside my present job. I'm
faced with a question of elimination."

"I'm not interested in your bowel problems," snapped
Cyprian, annoyed at being cut off short.

"You had opportunity and motive."

"Motive? I don't even know this bone-headed General of
yours."

"No. But as a leader of the guerrillas—"

"What on earth—?"

"As a general in this holy war against the establishment,
you would naturally want to kick off its most influential members. You've just told me that Stephen Protheroe is one of
them." Nigel had noticed that the sliding window was open;
he hoped Stephen was getting an earful. "Mr. Protheroe was in
charge of the proof copy. He had far the best opportunity to
do the dirty work, and was bound to be under the gravest suspicion. By stetting the libelous passages, Mr. Gleed, you knew
you had a good chance of losing him his job. Down goes one
of the pillars of the establishment."

After an amazed silence, Cyprian Gleed found his voice. "But that's absolutely—" He was about to say "fantastic"; but to say it would be to half admit that his idea of an Establishment was itself fantastic. "I really can't discuss this. I came here to get an apology from you for the insolent way you treated Miss Sanders."

"Your journey has been in vain, then. Good morning."

Trembling again, Cyprian scratched at his straggly beard. "Very well. I shall report your behavior to the partners."

Nigel, ignoring this, put a sheet of paper in the typewriter and began typing. He was pretty sure that Cyprian Gleed had not come here to take up the cudgels for Miss Sanders: Cyprian hardly seemed the chivalrous type. After biting his nails for a bit and glancing covertly at Nigel, the young man said,

"Perhaps I've been rather outspoken."

Nigel stopped typing, and looked at him silently.

"I've no objection to talking about your problem, actually."

I was right, thought Nigel: he came here to pump me. But on whose behalf? His mother's? His own? Cyprian now answered Nigel's questions with apparent frankness. Yes, he had come to Wenham & Geraldine's the morning of July 24th. Somewhere around 10:30, he thought. After a few words with Miriam, he had gone upstairs. Why? To see his mother. Why? Well, actually he'd run short of money and wanted an advance on his allowance; he'd failed to catch her before she left her house for the office. How did he remember this, when it had happened months ago? He had examined the stub of his mother's checkbook, which showed that it was on July 24th she had given him the advance.

"You were lucky," remarked Nigel. "I can't imagine many writers liking to be interrupted in the middle of their work with a request for money."

"Oh, that was part of my plan. When I came in, I started reading a page of her typescript and buttered her up about it," said Cyprian, with appalling candor. "She has abnormal vanity, even for a woman writer. And I don't usually take any interest in her work—it's entirely worthless, of course. So that did the trick."

"I see."

"When I'd got the check, I beat it—before she could change her mind. So, you see, I couldn't have done anything to that proof, even if I'd wanted to. I was in here with her all the time. I expect she'd remember, if you asked her."

"But you could have nipped into the next-door room before visiting your mother?"

"I could have. But it wouldn't have been any good—not if I'd wanted to fiddle with the proof."

"Because Mr. Protheroe was there?"

"No. Because my mother was there."

Betraying none of the emotion this statement roused in him, Nigel got up and closed the sliding window.

"How do you know?"

"I saw her through that little window."

"Saw her talking to Protheroe?"

"No. She was alone. She'd gone in to borrow an eraser. Actually, she was looking for it on his desk when I peeped through. I called to her, and she came straight back here. So she can give me an alibi." Nigel contemplated Cyprian so steadily that the young man added, with a roll of the eyes, "Honestly. You've only to ask her. Well, I suppose she mightn't remember, but—"

"I don't doubt you." Nigel continued to study him. Cyprian was clearly *capable de tout*; but did he know just what he was saying? "I don't doubt you. But I might doubt the eraser."

"What do you mean?"

"Your mother could have been tampering with the proof," said Nigel flatly. Cyprian Gleed received the suggestion with none of the moral indignation he had evinced over Miriam Sanders; indeed, there was something like a gleam in his eye— a vicious, calculating gleam he at once suppressed.

"I suppose she could. It hadn't occurred to me."

"Do you think she would?"

"Well, psychologically I wouldn't put it past her. Women of that type sometimes enjoy making mischief, you know. Or it can be the change of life."

"Who would she want to make mischief for?" asked Nigel, getting up and going to a far corner of the room to put this execrable young man out of range of his boot.

"Oh, just mischief in the abstract. Mischief for mischief's sake." Cyprian was evidently about to enlarge on this when the door was flung open and the subject of the discussion swept in.

"I told you not to come here again," she said harshly. "It's useless. You're not getting a penny." Miss Miles became aware of Nigel's presence. Her personality seemed to go out, then light up again in a quite different pattern, like an electric advertisement, all in an instant. "Oh, good morning, Mr. Strangeways. So sorry to interrupt you. I didn't know you knew each other."

"We didn't," said Cyprian. "Strangeways was interviewing me—"

"I'm so glad. Cyprian's the brains of the family. I know he'd do splendidly if someone gave him the sort of work that suits him—wouldn't you, dear boy?"

"Not that sort of interview," replied the dear boy sourly. "Strangeways is a detective."

Laying her handbag on the table, Millicent Miles sat down. Her son was still sprawled in the armchair, contemplating her with the look of one who perceives that a difficult sum is coming out right after all.

"A detective?" she said. "But how fascinating! I'd never have suspected it. And you've been grilling poor Cyprian?"

Young Gleed grinned. "We've been discussing mischief in the abstract—and the concrete, too, *Mother*." The last word shot out in a venomously satirical tone. "Do you need me any more, Strangeways?"

"No. Good morning."

Cyprian Gleed picked up his black sombrero. He gave his mother a veiled look from the doorway, blew her a kiss, and departed.

"I do wish he'd settle down to something. He's got plenty of talent, but I'm afraid he's become rather a drifter. I blame myself."

"Oh, surely not?"

"That's very sweet of you." She sighed, gave him an appealing, ingenuous look. "But he was the child of a broken marriage. It might have been different if I hadn't divorced his father. Though I had cause enough, God knows. But why should I burden you with my little troubles?"

Nigel imagined the dialogue of Millicent Miles' best-sellers must run on very much these lines. Their author proceeded: "You're investigating this business about Thor's book, I suppose?"

"Yes. Your son told me—"

"You're a very naughty man. You made me believe you were a new reader, and I'd hoped you'd persuade the firm to reprint those novels of mine."

"Afraid I can't help there. But you've a strong supporter in Mr. Ryle."

"Oh, Basil," she said offhandedly. "Yes, I suppose so. He's certainly a very persevering young man. But it's all so difficult." She sighed again. "I've often wondered if it's a good thing to have personal relationships with one's publishers. Perhaps it should be kept on a strictly business footing."

Nigel received this outrageous speech without comment. Could she be shocked out of her complacence, self-deception, whatever it was?

"Is your son an inveterate liar?"

"Really, Mr. Strangeways!"

"Can I rely on his evidence?"

"What evidence? I don't understand."

"He told me that he'd seen you in Mr. Protheroe's room

the morning the proof copy was altered—as good as caught
you in the act."

Nigel explained in detail. A sequence of emotions passed
over Miss Miles' face, but he could not tell how far they were
genuine; she might just have been trying them on and discarding them, like new hats. Her final selection was the
wounded mother's.

"Oh, Cyprian! How could you," she said brokenly. "What
an age we live in, when children bring evidence against their
parents. And false evidence too!"

"You weren't in Protheroe's room at all?"

"How can I possibly remember? It was months ago. And
I've often popped in there."

"Your son came to ask you for an advance on his allowance.
Does that recall it?"

"I'm afraid not," she replied, with mournful, hushed sweet
dignity. "You see, it's happened so often. Cyprian gets into
debt. I can't think what he spends his money on. Oh dear, did
he try to borrow from you?"

"No . . . You mustn't mind my asking this—but did you
ever see the libelous passages in *Time to Fight*?"

Her brow furrowed with concentration. "Let me think. Of
course I knew about them. Mr. Protheroe told me the sort of
thing Thor said. And I dare say I have seen the proof lying
about on his desk. But—"

"And you talked to your son about it?"

"I expect so." Her green eyes opened wide. "Oh, you can't
possibly think that Cyprian would do a thing like that?"

Her voice lacked conviction, and Nigel thought it was intended to do so. He felt a profound disgust; talking with emotional twisters, like Millicent Miles and her son, one became
tainted oneself. He was never nearer to throwing up the case.

"I'm trying to eliminate the people who couldn't have done
it," he said. "Physically, or psychologically. Have you known
Mr. Geraldine long?"

She gave her rattling, good-comrade laugh. "Now him you

can't suspect! I met him many years ago. Under rather strange circumstances. And then not again till recently. Why?"

"I'm groping in the dark—to find links. Is there any link between Mr. Geraldine and General Thoresby, for instance, apart from the publishing one?"

"I've no idea."

Millicent Miles was visibly losing interest. She had an inward look now, as though debating some course of action. Her face hardened.

"I must say that second libelous passage is marvelous stuff," said Nigel. He began quoting from it. " 'But the Governor, occupied with the more congenial business of cocktail parties . . . took no action whatsoever except to obstruct the military. . . . As a result of his criminal negligence, there was considerable loss of life and widespread destruction of property. The massacre at the Ulombo'—"

" 'Holocaust,' " interrupted Miss Miles in a distrait voice. " 'Massacre,' surely?"

"No, 'holocaust.' I've a first-rate verbal memory." She was giving her full attention now. "I particularly remember Stephen reading that bit out to me. It stuck in my mind—the word 'holocaust'—I've never been sure how to spell it. But my spelling's awful, anyway."

A few minutes later Nigel left her. He could not know that one of the questions he had asked this morning would lead directly to a murder.

VI *DELETE*

"**W**hat action would you take if I told you that it was Stephen Protheroe who tampered with the proof?"

"I shouldn't believe it," said Liz Wenham. "Apart from anything else, his whole life is bound up with this firm. It's quite inconceivable."

"But supposing I gave you absolute proof?" Nigel persisted.

"Well, we'd have to get rid of him," Arthur Geraldine said, after a pause. "Oh yes, Liz, we would. But surely you haven't—?"

"And he'd find it difficult to get another job?"

"Another job like this—yes."

"But, of course, leaving us would be by far the worst blow to him," said Liz. "I don't mean just financially."

"And we've agreed that he had infinitely more opportunity to do it than anyone else here?"

"Yes, but—" The two partners began speaking together, shocked by the enormity of Nigel's suggestion.

"No, wait a minute," he interrupted. "Stephen had no apparent motive whatsoever, you're saying. I agree. But perhaps somebody had a very strong motive for discrediting Stephen, ruining him. We've been thinking in terms of a culprit who wanted to damage either the firm or General Thoresby. But all we know so far is that Miss Miles was publicly humiliated by the General fifteen years ago, and that the firm turned down a project of her son's in July. As motives, these seem ludicrously inadequate."

"I agree."

"On the other hand, Stephen Protheroe was the main obstacle here to Miss Miles' having her novels reprinted. She was seen, alone, in his room during the crucial period. She knew about the libelous passages—as I told you, her using the word 'holocaust' that time is damned difficult to explain away except as a Freudian slip. She'd not been about the place long enough, in July, to know that you would never sack Protheroe on mere suspicion, however strong."

"But it's all so melodramatic and trivial," said Liz Wenham. "I don't particularly care for Millicent Miles, but I can't see her doing it just to get her novels reprinted."

"If that was her only motive, I'd agree. But I'm convinced she has some much deeper grudge against Stephen."

"How can she? They've only known each other for a few—"

"That's what they say. I don't believe it. They hate each other's guts. They talk to each other with a—the sort of familiarity bred by contempt, or hatred—as though they'd grown old in it."

Arthur Geraldine's sharklike mouth stretched in a grimace. "You may be right. But why should they pretend they'd never met each other before?"

"Why indeed? My head's buzzing with whys. Why, for instance, do Wenham & Geraldine allow Miss Miles to take up residence here indefinitely? She finished moving house months ago, after all."

The senior partner looked a little disconcerted. "It's a tradition of the firm that we should always keep a room available for our authors."

"The fact is, she's extremely difficult to dislodge," said Liz Wenham. "I've hinted often enough. You really must tell her quite definitely, Arthur. You were going to do so in September."

"For that matter," asked Nigel, "why should she want to stay on?"

"Meanness. Saves her burning her own electricity."

"Ah, come now, Liz, she's not as bad as that."

"Or to be on the spot when the proofs of *Time to Fight* came along, in the hope she could get at them somehow and do Protheroe down?" suggested Nigel. "Why else remain next-door neighbor to a man she abominates?"

They were in Mr. Geraldine's room, whither Nigel had been summoned, shortly after his conversation with Millicent Miles, to give a progress report. He had told them that hand-writing identification was impossible in this case, and finger-print tests would probably be useless. The proof had been handled legitimately by at least half a dozen members of the firm, apart from the author, the compositor and Nigel him-self. If the culprit was somebody who had no legitimate reason for handling the copy, he would presumably have taken care not to leave prints on it. Nigel was quite prepared, should the partners ask it, to fingerprint everyone in the firm; but Geraldine and Liz Wenham had agreed that, with so little chance of getting results, it would not be worth the trouble involved and the bad feeling among the staff that would ensue. It was decided, therefore, that only those who had been authorized to handle the proof should be asked privately to give Nigel a complete set of fingerprints. If some other prints were then to be found on the proof, the partners would decide what steps should be taken to identify them.

"I should enjoy being present, Arthur, when you asked Miss Miles for her dabs," said Liz, grinning amiably.

"God forbid it should come to that!"

The senior partner was clearly beginning to repent of ever starting this investigation. Engaged with the preliminaries to the libel action, he was looking preoccupied today, though his old-fashioned courtesy did not fail. He kept coming back to General Thoresby's highly unethical behavior.

"But look here, Strangeways. You say he as good as told you that he wanted to provoke a libel action. I never heard of such a thing! I shall instruct our lawyers to take it up."

"There was no other witness when he said it. He could deny saying it in court. It'd be my word against his."

"He seems to have the best motive of all for fiddling with the proof," said Liz.

"Yes. But he couldn't have done it himself. And can you imagine Protheroe, or Ryle, or the printer accepting a bribe from him to do it? Besides, he told me he was not guilty, and I believe him."

There was a brief silence; then Liz said, "So we come back to Miss Miles, or her son, or Stephen. . . . You've no other candidates?" she added, when Nigel did not reply.

Nigel had one other, but he was not prepared to talk about him yet. "No," he said. "Let's see if anything comes out of the fingerprint tests first."

"And if nothing comes?"

"Then I'd advise you to drop the investigation. I could spend days or weeks, and a great deal of your money, digging into the past to find some link between Protheroe and Miss Miles. But would it be worth your while?"

"I don't like digging things up. You never know what you may find," said Liz in her forthright way. "You might discover that Arthur had kept a brothel and Millicent was one of its inmates."

Geraldine's large pink face went pinker. "Really, Liz, you have an atrocious mind. Now, how do we set about this fingerprinting business?"

They discussed the procedure for a little; then Nigel went home to fetch his apparatus. After lunch he took the prints of the partners, Protheroe, and two members of the staff through whose hands the proof had passed. None of these raised any objection; nor did Mr. Bates, General Thoresby, and the manager of the printing works, when Liz Wenham explained to them over the telephone that it was to be done simply for the purpose of elimination. These, too, Nigel visited in turn; and by six o'clock he had taken complete sets of fingerprints from all concerned. It was understood that he would not be returning to the office till after the weekend, since the tedious task of bringing up latent prints on the 250-odd pages of *Time to Fight* and comparing them with the specimens he had obtained would take him the best part of two days.

He worked till midnight that day, with powdered graphite, insufflator and magnifying lens. He tested first the pages containing the libelous passages, which yielded him, amidst many blurred, indecipherable prints, a few he could identify as General Thoresby's, Stephen Protheroe's, Basil Ryle's and the compositor's. He worked all Friday, till teatime, at the thankless task, thinking sourly that this was the sort of routine job which ought to be done by the police. He had reached page 200 by four o'clock, and got nothing for his pains but a blinding headache and a conviction that he had a cold coming. He rang Clare Massinger.

"It's Nigel. I'm in a foul temper. Can I come to tea?"

"All right. Buy a Madeira cake on the way, will you?"

Five minutes later he was in Clare's studio. Gazing at a clay head on a stand, she did not look up for a few moments after he had entered: he had time to admire yet again the incredibly lustrous black hair that tumbled over her shoulders, the pale exquisite face in profile. Clare muttered malevolently at the head, then came over to Nigel.

"I won't kiss you," he said. "I think I'm getting a cold."

"I like your colds." Then, standing back from him, eying him with the same detached, concentrated look she gave her sitters, Clare said, "Yes, you do look rather disintegrated. I suppose you've been going about in wet socks again."

"Don't be a scold."

"You'd better have two aspirins with your tea. And take some nose drops in case your sinus starts playing up." She rummaged in a cupboard that appeared to be full of paint rags, producing a small green bottle with a syringe in it. "No, not that way. Lie down on the divan, hang your head backward over the edge like Desdemona after she's suffocated, and squirt it up your nostrils. The stuff won't get to your sinus unless you're upside down. No! Not a whole syringeful—just a few drops. Why won't you ever do things by halves?"

His head dangling over the divan's edge, Nigel reflected that there are few more agreeable experiences than being bullied by a beautiful woman. Sitting up again, he told her so.

"I'm not bullying," she said. "I just don't want to have you on my hands moaning about oh your poor head."

As Clare got the tea ready, Nigel looked round the studio. My home-from-home—and what a pigsty. I could never co-habit with a woman who keeps an open pot of foie gras on a shelf next to an open tin of turpentine. Cooking utensils, plaster casts, half-finished cigarettes, valuable books of re-productions, gobbets of clay were littered everywhere. A clay-stained smock hung on a hook beside a superb, violet-colored Dior coat. A shelf running the length of one wall held a herd of those small, stylized clay horses, with hoselike muzzles, which Clare molded almost automatically while her mind was at work on other things. On a pedestal in one corner stood her acknowledged masterpiece—the bronze head of a boy, peaky and perky, the lips shaped as if he was about to whistle a tune or utter some blistering and unprintable repartee. It was the head of the boy Foxy, who had come into their lives so un-ceremoniously sixteen months ago, during the events which had led up to a melodramatic scene in the Albert Hall.

"Foxy wears well."

"Yes. He dropped in to tea yesterday. Asked after the Guvnor. I told him you were busy detecting. Oh hell, I wish I could detect the cake knife."

"It's in the second drawer of your make-up table, in the bedroom."

"*Is* it? How did it get there? . . . Oh, you're mocking me!"

Clare, who had been prowling the studio in search of the cake knife, swirled round, and accidentally jogged Nigel's tea-cup out of his hand.

"Oh damn it all, *why* am I so clumsy!" she wailed, picking

up the cup and mopping Nigel with the scarf she had taken off her neck.

"You're not. You're extremely graceful. It's just your impulsive nature coming out that makes you bump into things."

After tea, Nigel lay on the divan with his head in Clare's lap. Her fingers stroked and molded his temples; he felt like a piece of clay being brought to life. Her hands were square, stubby-fingered, strong but also meltingly soft. Nigel kissed one of them.

"They're ugly, aren't they?" she said. "That's why I take so much trouble over them."

"I wish I was ugly, then."

"Oh, you are. Hideous. In an interesting way. Your face is lopsided. But symmetry is boring. Do you feel better now?"

"Much better."

"Sometimes I wonder why I don't marry you," said Clare dreamily. "At any rate, it's not for want of not being asked."

"Or not asking."

"Well, will you?"

"I couldn't live in this sort of chaos, darling Clare. I should go mad."

Her dark eyes widened. "But I'm not asking you to *cohabit* with me. I don't think that would suit me at all. I wouldn't want you about the place all the time, tidying things up. I said 'marry.' You see, I like my chaos."

"Chaos is a necessary condition of creation for you?"

Clare nodded, and the coal-black hair danced over his face. "Pompous old bastard," she lovingly murmured.

"Shall I tell you all about my case?" said Nigel presently.

"I should like that very much."

Nigel removed himself to an armchair, and having shifted from its seat a large ammonite, an ash tray and a half-eaten bun, sat down. He then told her, in very considerable detail, about the libel affair at Wenham & Geraldine's. Clare listened intently, curled up like a cat on the divan, her eyes regarding him unwinkingly, like a cat's.

"Well, I call that very interesting, in a quiet way," she said when he had finished. "I don't like the sound of that Basil Ryle much."

Nigel glanced at her alertly. It was odd she should have picked out the suspect he had not mentioned to the other partners this morning.

"I think he is riding for a fall," she continued.

"Why?"

"You say he's in love with this Miles woman. I happen to know that she's bitch-bitch number one."

"You know her?"

"Know of. The Miles woman has one constant, enduring passion, one soulmate—Millicent Miles. She will have been keeping this Ryle on a string. If he fails to get the firm to reprint her books, she will cut the string. If he succeeds, she'll have got what she wants out of him, so she'll let go of the string. Either way, he's had it, you mark my words."

"Well, I must say, Clare, sometimes you amaze me."

"What? Am I wrong?"

"No, I'm sure you're right."

"And this Ryle sounds to me an explosive character. I suppose he's repressed and old-fashioned under his chromium plating, and nourishes rather shoddy-glamorous-chivalrous notions about sex and is half annoyed at himself for it?"

"How long have you known him?"

"Never met him in my life, or heard about him till you told me. Honestly."

Nigel rose and started to pace the studio. "Suppose Millicent Miles has made it a condition of marrying or going to bed with him that her books get reprinted. Protheroe is the main obstacle to this. So Ryle has an exceedingly strong motive for discrediting Protheroe, quite apart from his general feeling that Stephen's a back number—like Bates, the late Production Manager—and ought to get the push."

"Oh, I don't think Ryle did it," said Clare.

"Why not?"

"He's been in the firm long enough to know that this Protheroe was the pomme of their yeux. They'd never sack him just on suspicion. No, I think Protheroe did it."

"But really, Clare—"

"Yes, I know. Fouling his own nest, and all that. But he sounds a bit mad, from your description. Anyone would be, reading books all day for twenty-five years. You'd lose touch with what they call reality, wouldn't you? I expect M. Miles egged him on to do it. All this quarreling they put on in public—it doesn't ring true to me, except as camouflage."

"Why should she egg him on?"

"Or perhaps she blackmailed him into it."

"Why?"

"To spite the General. Or the other General—the one who's bringing the libel action. I'm sure she's a vindictive woman. Perhaps the other General—"

"Blair-Chatterley."

"—perhaps he's the man who seduced her when she was a pure young maiden, if ever she was, which I very much doubt: the mystery man Ryle mentioned to you. So when she came out with that word 'holocaust,' she and Protheroe were really gloating. Perhaps the child she had by Blair-Chatterley was one of the chaps who perished in the holocaust. Revenge motive, see?"

"It was stillborn. You're raving, my dear girl. But as you're interested, you'd better do some research for me. You said you know of Miss Miles—"

"Yes. A woman who sat for me last year knew her—some time ago, though."

"The longer ago the better. Find out from this woman—no, why not go straight to the horse's mouth? Invite Millicent Miles to sit for you. She's vain enough to accept."

"Oh, all right. If I must," said Clare dubiously.

But, as it turned out, the circumstances called for a death mask of Millicent Miles, not a portrait head. At some time this very evening, perhaps while Nigel and Clare talked about her, she was having her throat cut.

Millicent Miles was at work on what was destined to be an unfinished autobiography, when it happened. The blinds were down, the electric fire was burning, the typewriter rattled away in the room at Wenham & Geraldine's, and on the table beside it the pile of typescript silted higher. The frosted-glass window between this room and Stephen Protheroe's was closed, its square pane showing no light from the other side. Millicent glanced at it for a moment, screwing up her eyes against the smoke from her cigarette; she glanced at her watch, then fell to work again. Whatever her other faults, as a writer and as a human being, she had remarkable power of concentration—so remarkable, indeed, that when the door behind her opened she did not look round. Either she was too absorbed to hear the door open, or she expected a visitor.

The visitor, in a swift movement, closed the door and turned the key. He then—let us say "he," though for reasons which will become apparent later, the sex of the visitor might be open to question—he then laid a grip bag on the floor, and in the same action took a long pace which brought him directly behind Miss Miles' chair.

"Just a minute," she said, still not looking round. As last words, these were both undistinguished and unavailing. The visitor was not prepared to wait even a minute. His gloved right hand thrust round over her arms, which were stretched out toward the typewriter; and his right arm, clamping them to her body, tilted the chair back. At the same instant his gloved left hand stuffed a piece of cloth into the mouth as it opened to cry out, knocking the cigarette away in the process. The visitor then swiftly changed arms. His left one went round her arms and body, drawing the chair away from the desk and tilting it back still further, while his right hand dived into a pocket, whipped out a razor and opened it. Her staring eyes hardly had time to change from astonishment to terror before the razor had done its work.

All this had taken no more than ten seconds. The visitor let the back of the chair gently down to the ground, took the dying woman under the armpits and dragged her body off the chair to a corner of the room, where he let it lie flat. She was still making gurgling sounds, so he thrust the cloth deeper into her mouth and throat. He then—and throughout, the visitor performed with the swiftness and precision of an actor who has rehearsed his stage moves to perfection—took a staple from his pocket, and with a heavy ebony ruler he had picked up from the table hammered one point into the frame of the sliding window, the other into its wooden surround. Four firm blows, and the window was fastened shut.

Next, the visitor removed his bloodstained gauntlet gloves; he took a clean pair from the bag, put them on, and sat down to the typewriter. Flicking out the sheet which was already there, he inserted a fresh one and began to type. A few minutes later, a door banged down the passage. The visitor took a deep breath, but continued typing as the feet approached the door. They passed it, hurrying. If there were other people moving about in the building further off, the typewriter drowned any noises they might be making.

The visitor now took the sheet out of the machine, and turned to the pile of typescript on the table. Though his gloved hands made it difficult to sift the pages, he soon found what he wanted. Slipping out a certain page of Miss Miles' autobiography and crumpling it into his pocket, he substituted for it the one he had just written. He moved to get up, but as if on an afterthought glanced over a few sheets of the typescript which came before the one he had removed. Something caught his eye, making him draw in his breath sharply. He reached for the eraser, examined it to see there was no blood on it, then carefully rubbed out whatever it was that had offended him. After looking through several more pages, he placed the pile of typescript face down on the table again.

He now lifted the typewriter and placed it on the floor, choosing a dry spot, near the body of Millicent Miles, who by this time was evidently dead. Raising each of her hands in turn, and wiping first the several fingers which had blood on them, he pressed the fingertips against the keys of the machine, then replaced it on the table, and reinserted the sheet he had originally taken out.

Throat-cutting is a messy procedure. There was a great deal of blood on Miss Miles' dress, and on the floor beside her, and in the middle of the room where she had been murdered. There was blood, too, on the forearms of the visitor's coat, and a few stains elsewhere received from its contact with the chair. But the visitor, who had otherwise manifested so much forethought and care, did not take any great precaution against stepping in the pools of blood. He gave a last look round the room—a look which comprehended everything except the hideous gash in the victim's throat. He tapped his pocket: the razor was there all right. He took off the galoshes he was wearing, wrapped them in newspaper and put them into the bag. Finally, moving round the edge of the room where there was no blood, he levered out the staple from the sliding window and returned to the door.

For a moment he stood there, listening intently; glanced at his watch—he had been in the room for a little more than fifteen minutes. Taking up the bag, which appeared to weigh heavy for its size, he turned out the light, unlocked the door, and was in the passage. Locking the door from the outside and pocketing the key, the visitor walked down the dim-lit passage toward the lift.

VII *QUERY*

Nigel Strangeways arrived at Angel Street at 9:30 on the Monday morning, his cold in full flood. He had discovered no unauthorized fingerprints on the proof of *Time to Fight*; and however fascinating might be an investigation into Millicent Miles' past, he felt sure the partners would not spend money on it. The case was over, for him. The trail had gone cold months ago, and it would never be possible to prove who had let Wenham & Geraldine in for this disastrous libel action, however strong his suspicions might be. Of material evidence there was none.

Entering the reception room, he noticed Miriam Sanders looking worried.

"What's the matter?" he asked.

Worry melted her hostility for a moment. "It's the key," she replied, glancing down at her desk and up again.

"What key?"

"The key of the room Miss Miles is using. The cleaners left a message to say they couldn't get in there this morning. The door was locked."

"Well, I expect she took it home by mistake. Why not ring her up?"

"I did. But she's not at home. The German maid sounded in rather a state—I couldn't make out what about—she's got hardly a word of English."

"It's not your worry."

"But Miss Miles is not supposed to take the key away without telling me."

Nigel dismissed the matter and went upstairs, wondering vaguely why Miss Sanders was fussing. Stephen Protheroe was already at his desk. They discussed their weekends—Nigel nursing a cold at home, Stephen visiting friends in Hampshire —for a few minutes. Arthur Geraldine came in.

"Did Miss Miles say if she was coming back today?"

Protheroe shook his head.

"Apparently she's locked her room and taken away the key. It's a damned nuisance. I told her on Friday morning we needed the room for other purposes."

Stephen sniffed. "Typical! If she can't have the room herself, she'll see to it that we can't use it."

"But surely there's a spare key?" Nigel asked.

"It's been mislaid. I've sent for a locksmith to open the door."

"Well, that's all right then, isn't it?"

But Arthur Geraldine remained, oddly irresolute, gazing out of the window, then at a calendar on the wall, till Stephen remarked irritably, "If you want to know whether she's coming back, look and see if she's left her typescript on the table."

"But we can't get in, I've just told you. . . . Oh, I'd forgotten the sliding window."

Arthur Geraldine, emanating a queer uneasiness, grasped the knob resolutely, as if it were a nettle, and after a moment's hesitation pulled the window open.

"Yes, the typescript's— My God, she's there! On the floor!"

Nigel had to shoulder Geraldine away from the window— he seemed frozen to the spectacle within. Though the blinds were down in that room, enough light came through them and the sliding window to show the body lying in the far left-hand corner: Nigel made way for Stephen Protheroe, who exclaimed, his voice rising to a squawk,

"She's cut her throat!"

Arthur Geraldine was muttering in a distraught way, "I

knew there was something wrong, I knew there was something wrong." His face trembled like a pink jelly.

"How could you possibly know?" Stephen sounded irritable again.

"I've been feeling uneasy all the morning. Must be my Irish blood."

Stephen was now positively exasperated. "I suppose you heard a banshee in the night too."

Nigel was at the telephone. Presently he put down the receiver. "Scotland Yard are sending a team and a police doctor. Mr. Geraldine, will you instruct someone to be waiting for them. And tell Miss Wenham what has happened. I'll tell Ryle myself."

"But we can't just leave the poor creature there."

"No one must enter that room till the police come. She's dead. There's nothing you can do about her."

Geraldine plunged out of the room. Nigel rang Basil Ryle on the internal telephone. "Would you come up to Protheroe's room? Yes, it's urgent." Nigel turned to Stephen. "Let me do the announcing, please."

Basil Ryle looked strained and gummy-eyed, as if he had not slept. His voice had the sandpaper rasp of exhaustion. "What's the matter now?" he asked.

Nigel motioned him to the sliding window. With a puzzled expression, Ryle approached it and looked through. He was quite still for a moment; then his head bowed down slowly till the forehead rested on the window ledge, as if in prayer.

"No! No! Millicent! No! No!" The almost inaudible gabble trailed away, and Basil Ryle slipped sideways to the floor.

"Well, I must say," Stephen began to protest.

Nigel glanced up from loosening Ryle's tie and collar. "What time did you leave here on Friday?"

Protheroe's face seemed to diminish; the lips pushed in and out.

"Are you completely inhuman?" he said.

"It's the first question the police will ask you. Keep the answer for them, if you prefer."

"Soon after 5:15, I think. I had a train to catch."

"And was Miss Miles in there when you left?"

"Yes. Typing her book, though, not cutting her throat."

Basil Ryle moaned, as if in comment on Stephen's edgy remark; his eyes opened and he sat up, shaking his head. "What on earth? . . . Did I faint?" Then, as full consciousness returned, his face settled back into an expression of strained, drained misery, and he struggled to get up.

"It's no good, boy," said Stephen with unusual gentleness. "You can't do anything. She's dead."

Ryle stared at him. "I'm sorry," he said at last; and it was likely that he referred, not to Millicent's death, but to his own past attitude toward Stephen Protheroe.

"There *is* something you could do," said Nigel. "Run downstairs and ask Miriam Sanders for the key—"

"But the key's gone," Stephen said. "No, of course, she must have locked herself in. But—"

"If you'll look through that window again, you'll see that the key's not in the door. Of course, she might have locked the door and put the key somewhere else in the room. But I didn't mean *that* key."

"Have you any idea what you do mean?"

"The spare key of the side door into the street. The receptionist keeps it. I'd like to see it, please."

Basil Ryle went off on his errand. Nigel anticipated another protest from Stephen: "No. It's best to give him something to do. The office is closed from Friday evening to Monday morning?"

"Yes."

"No one here at all? Ever?"

"No. Well, of course Arthur has a flat on the top floor. You know that."

"And what about Friday evenings? Staff leave at the same time as other weekdays?"

"Yes. Some at five, the rest at five-thirty."

"Except those doing overtime?"

"No. We've a strict rule there should be no overtime on Friday evenings. The partners often leave earlier than usual on Fridays too."

The telephone rang. It was Basil Ryle. Miss Sanders could not find the key of the side door. It was not in the drawer where she normally kept it, nor in any other drawer.

"Ask her when she remembers seeing it last."

"Thursday morning, apparently."

"Thank you. You didn't borrow it yourself? No. Would you please ask the other partners if either of them did?" Nigel put down the receiver. "That's something else missing. Probably in Miss Miles' handbag, though."

"Something *else*?" asked Protheroe.

"Yes. The other thing is the razor."

"The razor?"

"Or whatever she cut her throat with. It's not in her hand. And I don't see it lying beside her. Ah well, we shall know soon enough."

Stephen's fine eyes regarded Nigel steadily. "You mean, whether it was suicide?"

"Exactly."

Nigel took up the telephone again, and had himself put through to Millicent Miles' house. After a conversation with her maid in German, he told Protheroe, "She didn't go home on Friday evening, though she'd ordered dinner. The maid was not particularly surprised—her employer often altered arrangements without letting her know: she assumed Miss Miles had gone off for the weekend. She tried to get in touch with Cyprian Gleed, on Sunday, but failed."

"So you think it happened on Friday evening?"

"Looks like it. But there's no knowing yet."

A tramping of feet on the staircase. Nigel went out. Arthur Geraldine was leading a party of C.I.D. men headed by the saturnine Inspector Wright, who had recently been transferred from Division to H.Q., and the police surgeon. Wright raised his eyebrows at Nigel, but gave no other sign of recognition. Geraldine made the introductions.

"I expect you'll want a room to—er—work in. Interviews and so forth," he said. "You'd better have the reference library. Second door on the right."

"That's very good of you, sir. I won't keep you from your work any longer, then." Wright glanced at his Detective Sergeant, pointed at Miss Miles' door, and made a turning movement with his other hand. The man took out a bunch of skeleton keys.

"I wonder could you spare us a few minutes downstairs, Mr. Strangeways?" Geraldine's courtly manner struck a queer note under the circumstances. "If there's anything else you should require, Inspector, ring extension 4. That's my number."

Nigel waited behind for a word with Wright. As they talked, the Sergeant got the door of the room open. A breath of very warm, unpleasantly stuffy air came out. The electric fire had been burning there all the weekend.

Downstairs, in the senior partner's room, Nigel found a replica of the scene which had met his eyes on his first visit, five days ago: Geraldine at the desk, Liz Wenham leaning against the window, and Basil Ryle jingling the coins in his pocket on the other side of the desk. They might have been arrested eternally in a conversation-piece grouping—three publishers discussing the spring list or a reprint order or some author's witless objections to his blurb.

"This is a terrible business, Strangeways," said Arthur Geraldine. "Why should she do it? Why?" And within the sacred precincts of Wenham & Geraldine, his scandalized face visibly added.

"It's by no means certain she did do it."

At the window, standing very still, her rosy-apple cheeks looking blotched now, Liz Wenham said, "You mean, it

could have been—" the word "murder" was too much for
her—"could have been done by somebody else?"

Nigel nodded. He noticed Ryle flinching. Geraldine buried
his bald head in his hands.

"First the libel trouble, now this," he muttered.

"Oh, come, Arthur. That's just coincidence." Liz Wenham
was trying to be her brisk self. Coincidence, thought Nigel:
what about coincidence now?

"She'd never do it. Not Millicent." Basil Ryle's tone was
a strange mixture of agony and exasperation. Liz gave him a
disapproving glance. She is shocked by naked emotion,
thought Nigel; in business hours, at any rate; keeps her life
in compartments. He realized he had no conception what
went on in her other, private compartment.

"I suppose we ought to get in touch with her relatives," Liz
was saying. "Is Cyprian Gleed the next of kin?"

The telephone rang. Geraldine gave them a rueful look, and
became involved in a conversation with the caller about a print
order. Drawing Nigel aside, Liz Wenham said, "We must
have your help over this. Your professional services, I mean.
Will you stay on for a while?"

"Of course, if you want me to. But I'm afraid we're at a
dead end with the libel business. Unless it's connected—"

"Basil. Do you know Gleed's number? Ring him up, like a
good chap," said Liz.

Ryle stared at her intently, as if trying to understand a for-
eigner, then walked out of the room like an automaton. A
secretary came in with a message for Liz.

"Tell her I've got to cancel lunch today. I'll ring her back.
Oh, and, Laura, ring Clausson and tell him he's not got that
second color right yet on the Bellington jacket: it's still too
muddy."

"Yes, Miss Wenham." The girl tiptoed away, solemn as a
communicant. Rumor of Miss Miles' decease was already
going the rounds.

Liz Wenham pushed back the gray hair over her temple.

"Business as usual. Or do you think we ought to close down the office for today?"

"No. The police will want to interview your staff."

Arthur Geraldine laid down the receiver, took out a large silk handkerchief and mopped his face.

"What were we saying?"

"I've asked Basil to ring Miss Miles' son. And Mr. Strangeways is going to stand by for a bit."

"Oh, good. I hope Basil will be tactful. It'll be a terrible shock for—"

"Shock treatment is what that young man needs. Let in a bit of reality. Ruined by his mother, of course."

"Ah, come now, Liz."

"Don't be sanctimonious, Arthur. She was a poisonous woman, and you know it."

Geraldine gave her a strange look, then plied his handkerchief again as if to wipe the expression off his face. "Extraordinary the way one's mind works. I can't help thinking how this will put up the advance sales for the Miles autobiography. I'm afraid you must be getting a very low view of publishers, Strangeways. We're all monomaniacs, you know."

Nigel's polite murmur was drowned by an edgy laugh from Liz Wenham. "But even a publisher won't arrange an author's death just for publicity purposes, so you needn't look so anxious, Arthur."

"Liz, that's no way to be talking. You don't seem to realize—"

The door opened and Basil Ryle came in. "Well, I've told him. He sounded as if I'd waked him from a weekend hangover."

"Is he coming along?"

"I said there was no need to yet."

"Did you ask him when he'd seen his mother last?" inquired Nigel.

"Seen his mother last?" Ryle vaguely repeated, his eyes wincing. "No. No, why should I? I'm not a policeman."

"Well, there's no use standing about talking." Liz Wenham's voice was brisk. "I'm going to do some work. We'd better get down to that leaflet for the Hosking memoirs, Basil."

"For God's sake!"

"Come along. We can't have you going broody." There was compassion beneath the roughness of her tone. Basil followed her out meekly, his feet dragging.

"I've got to ring our solicitors," said Geraldine. "Any news?"

"No. I'm afraid it's a washout."

"Well, I'm sure you've done your best."

"I daresay if we put in some heavy research—"

"Research?"

"Yes. I've a hunch that the secret of this libel business could be found, if we dug deep enough. We'd find it buried somewhere in the past."

Arthur Geraldine raised his hand from the wrist, where it lay on the desk. "I don't think we'd better go stirring up any more mud, Strangeways. Not in view of what has just happened." The senior partner's voice was courtly but firm—the voice of Wenham & Geraldine regretfully declining to make an offer for an author's manuscript.

Nigel went upstairs to Stephen Protheroe's room. Stephen was concentrated upon a typescript, undistracted apparently by the voices and shufflings in the next room, the flash bulbs that lit up the frosted glass of the sliding window from time to time.

"How long does this go on for?" he asked, without looking up.

"All depends. A few hours, at least."

"And then?"

"They'll take the body away for the post-mortem and start interviewing everyone."

Stephen made a pencil mark on the page before him. Then, in a resonant, thrilling voice that came oddly from so minnowy a man, said:

"The bustle in a house
The morning after death
Is solemnest of industries
Enacted upon earth."

"Yes," said Nigel. "But it goes on about 'putting love away,' doesn't it? Not so appropriate."

After a pause, Stephen murmured, as if talking to himself, "What was her attraction? Horse teeth. A laugh like a football fan's rattle. Mouth too big, heart too small. Yet she could get any man she wanted—any man. I suppose all-of-a-piecenesss was her strength: the seamless garment of egotism; like a child's egotism—yes, there was something innocent about it. And innocence can be the most unscrupulous, destructive thing in the world. Of course, she had vitality too. Incredible vitality. That's the flame every moth goes for."

"I imagine her as rather a tomboy when she was young."

"Tomboy? M'm, yes, you may be right."

"Which doesn't fit in with her own picture of herself as a downtrodden sensitive at the mercy of a drunken father and a sluttish mother."

"What? Did she tell you that? No—" Stephen looked suddenly alive with intelligence—"no, that's the sort of romance she'd spin to catch some chivalrous, unsophisticated youngster, God help him."

"Her parents weren't—?"

"For all I know, they may have been fiends in human form. But I bet she gives a rather different account of them in her autobiography. Of course, she knew her market. The Millicent Miles fans would be terribly shocked if she undressed her dead parents in public."

"You've read the book?"

"I was tempted to take a peek at it, compare the actual Miles with the so-to-speak presentation copy. But I— What the hell—?"

Protheroe gave a start as the sliding window was flung open

and Inspector Wright's sallow, sharp face looked through. Nigel introduced them.

"You normally occupy this room, sir?"

"Yes."

"Can you tell me when this window was nailed up?"

"Nailed up? How do you mean? It's never been nailed up as far as I know." Stephen sounded quite annoyed.

"When did you last have it open, sir—before today, I mean?"

"Good lord, how should I remember? Friday afternoon, probably. Yes, I think Miss Miles opened it to have a word with me sometime in the afternoon."

"And, of course, you'd have noticed the sound of hammering, if it'd been done while you were still here."

"I imagine so."

"And you left the office on Friday at—?"

"Soon after 5:15."

"I'm much obliged to you, sir." Inspector Wright gave a little sideways jerk of his head at Nigel, who at once went into the next room. It seemed to be filled with bodies, apart from the one lying on the floor, covered now with a macintosh—plain-clothes men, each going about his task, separate and preoccupied, like people playing treasure hunt at a party. The blinds were up, and the window open: it was difficult to keep one's eyes off the blood patches on the floor, glistening and rust red.

Inspector Wright drew a forefinger across his throat, from ear to ear. "Murder. No weapon. Door key gone. We'll have a talk presently, Mr. Strangeways. Just take a look at these marks; they're recently made." He indicated two holes in the frame and surround of the sliding window. "Staple driven in to fasten the window shut, looks like. Why?"

Nigel was not unaccustomed to Wright's habit of keeping his subordinates on their toes in this pedagogic manner—asking them to explain what was already clear to him.

"To prevent anyone looking in while the murder was being committed," he replied.

"Yes?" The Inspector's dark, piercing eyes were still expectant; his fingers beat a tattoo on the air.

"Which suggests," Nigel equably continued, "that the murder was committed *either* on Friday evening, after Protheroe had left but while there could still be other people in the building, or was committed during the weekend and the staple driven in to make you think it had been done on Friday evening."

"Not bad. But have you considered why the staple should have been removed at all after it had served its purpose?"

"You tell me—it's your turn, mate."

"Assuming the murderer assumes we would not notice the marks it left, or wouldn't bother about them, it'd give him a pretty alibi. Who'd ever kill the woman, with other people still in the building, liable to walk into the next room and take a look through the window? That's how he wanted us to think. Therefore, we're meant to say, it must have been done over the weekend, which no doubt he spent a long way from here. Right?"

"Possible. But there could be another explanation. How long d'you take to walk into this room and cut a woman's throat? A few seconds, if you take her by surprise. You don't, presumably, staple the window first. But why do it afterward, unless you want to stay in the room for a while unobserved? And why should you want to do that, unless you're going to get rid of some real evidence or plant some fake evidence?"

Wright nodded vigorously. "O.K. in theory. But all we've found so far is a few footprints in the bloodstains. Galoshes probably. Size 10. Quite smart." He made a flapping motion with one hand. "Wear a pair several sizes too big. Fox the police—poor dumb clucks. Then whip 'em off. They'd wash easily too."

"When did she die?"

"You know these medicos. Not less than thirty-six hours

ago. Not more than three days. Helpful. Wait for the P.M., my lad."

"She didn't go home on Friday evening. And she hasn't been there since. I—"

"Keep it, Mr. Strangeways. We'll have a proper chat soon. Going back to those staple marks— Yes, what is it, Summers?"

The Inspector was drawn aside by one of his men. Nigel took the opportunity to inspect the table where the dead woman had worked. The pile of typescript was still there, and a sheet in the machine. Millicent Miles' career had been ended, it seemed, in mid-sentence. Nigel bent closer, something catching his eye.

"Wright, have you finished with this machine?"

"It's all yours."

The door opened. Men came in with a stretcher. After the body had been lifted onto it and carried down to the ambulance, Nigel did a little typing; then he beckoned to the Inspector.

"See? That's where she left off typing. And that's where I began. It's out of alignment."

Wright's eyes sparked with intelligence. "Someone took out the sheet of paper, then put it back later? She could have done it herself, mind you."

"She *could* have. If she didn't, the murderer had some typing to do himself. That's what he needed time for. That's why he stapled the window. Now *why* did he have to use a typewriter?"

VIII *LOWER CASE*

Late that night, on Nigel's invitation, Inspector Wright was supping with him at Boulestin's. It was a silent meal, for both men were reading. Nigel had spent the afternoon, while Wright and his detective-sergeant were interviewing members of the firm, in typing out a full account of his investigation of the libel affair. It was this that the Inspector was now reading, as he wolfed a delicate saddle of hare and Nigel studied the results of the police interviews.

From the latter, it was evident that Wright had thrown his net over the period between 4 P.M. and midnight on Friday. At about four, according to the girl on the switchboard, Miss Miles had put through a telephone call; the police had checked this call—it was to her hairdresser. Stephen Protheroe was vague as to when she had opened the window for a word with him—thought it would have been about 4:30 but couldn't be sure; so it was safer to assume 4 P.M. as the last time when she was known to be alive. Inquiry into Miss Miles' habits showed that she normally left Wenham & Geraldine at about 6 P.M.; but occasionally, according to her German maid, she had worked later, not returning home sometimes till 8 o'clock. However, unless she had made an appointment with the murderer for a meeting in the office late that night (and why at the office if it was to be so late?), the natural deduction was that she must have been killed before 6 P.M. The murderer would hardly have based his plans on the chance that he would find her still at work later than this. Besides, the clue

of the staple marks pointed to a period when there were still
people in the office and the murderer might be spotted with
the body.

There was one other time pointer in Inspector Wright's
report. He had interviewed Susan, the forward blonde of the
Reference Library. Susan had heard no suspicious sounds from
the next-door room that Friday afternoon; leaving at 5:30
sharp, she had heard "Miss Miles typing away like mad," as
she passed her door. Now this meant either that Miss Miles
was still alive then, or that she had just been killed and it was
the murderer whom Susan heard typing. Wright's investigation concentrated, therefore, on the period from 5:15 to 6
P.M. Inevitably, given these facts, it must; but to concentrate
thus, Nigel reflected, was to ignore the problem of the missing
key—the spare key to the side door, which had last been seen
by Miriam Sanders on Thursday morning. This key was not
found in the dead woman's possession. It must have been
"borrowed" by somebody else. Why borrow it except to get
into the office after the main door was locked at 5:30? The
partners and Stephen Protheroe each possessed a side-door
key, so could have no reason for abstracting the spare one
except to divert suspicion onto an outsider or some member
of the staff.

Withdrawing his mind from these speculations, Nigel applied it again to the evidence. The movements of the people
in whom he was particularly interested were briefly as follows:

Arthur Geraldine. 4–5:50, in his room; 5–5:15, dictating letters
(confirmed by secretary); 5:15–5:18, brief chat with S. Protheroe;
5:50, went upstairs to flat (confirmed by Mrs. Geraldine, who
said that her husband was with her for the rest of the evening).
Elizabeth Wenham. 4–5:15, in her room; 5:10–5:15, discussing
business with Basil Ryle; 5:15–5:20, in studio (confirmed); 5:20–
5:30, in her room; 5:30, left office (confirmed by M. Sanders);
6:00, cocktail party, Chelsea (confirmed); 7:30, arrived home
(confirmed by maid); dined alone, read a book, to bed at 11.
Basil Ryle. 4–5, out of office, giving talk at Book Fair; 5:10, re-

turned to office (confirmed by M. Sanders); 5:10–5:15, discussing business with E. Wenham; 5:15–6:00, working in room (confirmed by secretary for period 5:15–5:25); 6:00, to Festival Hall for dinner, then concert (not yet confirmed); 10:25, returned home.

Stephen Protheroe. 4–5:15, reading; 5:15, a few minutes' chat with A. Geraldine (confirmed), then left building at 5:20 (confirmed by M. Sanders) and walked to Waterloo Station via Hungerford Bridge; caught 6:05 Southern Electric to Pennshill, Hants; 7:30, met by friends at Pennshill Station (not yet confirmed).

The immense task of interviewing the Wenham & Geraldine staff individually had, of course, not been covered today. After he had interviewed the principals, Inspector Wright and his sergeant, conducted by Basil Ryle, went the rounds of each department, asking the employees only if they had seen or heard anything suspicious on Friday evening. The result was a complete blank. Miriam Sanders declared that no unauthorized visitor had come to the building between 4 P.M. and 5:30, when she locked up. A considerable number of the employees—those who clocked out at 5 P.M.—could probably be dismissed from the reckoning; but there still remained some thirty of them, apart from the principals, whose movements would have to be given a routine check.

Police were still searching the building for the weapon and any bloodstained clothes that might have been hidden there. The murderer must have got a good deal of blood over himself, and it seemed unlikely he would walk out of the office in such a condition.

The weapon, thought Nigel. A razor. Which of the people who had a motive for killing her would use a razor? Such a primitive sort of weapon, so unlike the highly civilized persons who controlled the fortunes of Wenham & Geraldine. Of course, there's always Cyprian Gleed. He found he had said it aloud.

"Yes," repeated Inspector Wright, looking up from Nigel's report, "there's always him. I sent a man to interview him this afternoon. Seems this Gleed was alone in his flat from 4:30 till 7 on Friday, then went out on a bender. Lives by himself over a shop in W.8, private door on street; no corroboration. Not very co-operative, my chap reported." The Inspector sketched an epicene gesture. "Cyprian Gleed! Sounds as if he'd walked out of Wilde or Firbank. Miss Miles, now—was she a touch typist, d'you know, or the old two-finger-exercise stuff?"

"How you dart about! I never saw her at work; but it sounded rapid and fluent enough to be touch typing. Why?"

"Just got a report from the fingerprint boys before I came along here. Only her prints on that typewriter. Prints of all ten fingers on the keys." Inspector Wright danced his own fingers on the table, glancing cannily at Nigel.

"But?"

"But the murderer was not conversant with the touch-typing system."

"Sorry, my head's not too good just now."

"The fingerprints are on the wrong keys, if you get me. He must have pressed her fingers on them regardless, after she was dead. Cold-blooded character."

"And after he'd finished his own bit of typing. It's another proof that he did some, isn't it? And that he doesn't want us to know he did?"

"Looks like it."

"Of course, she might have invented a touch-typing system of her own."

"Love making difficulties, don't you? The Oxford mentality. Cheers." Wright drained his glass of Richebourg, and began collecting his papers. "I'm going to sort out young Cyprian tomorrow morning. You'd better come as my bodyguard. I'll pick you up at 9:15. And thank you for my nice dinner."

Nigel's first cigarette the next morning, after an early breakfast, started a paroxysm of coughing which set his sinus aching fiercely again. He took some nose drops and put the bottle in his pocket. Last time he had worked with Inspector Wright, he had been coshed silly; and now this stabbing pain over one eye. Wright is my jinx, he thought sourly. The prospect of reading through Miss Miles' autobiography, after the finger-print experts had dealt with it, appalled him. The secret of the crime very likely lay imbedded somewhere in its perfumed pages; but the possibility roused no enthusiasm in him. All he wanted was to crawl round to Clare's studio and go to sleep.

In the police car, he began to feel a little better. Inspector Wright's vitality was infectious: he could not imagine any state less active than his own, and this keyed you up when you were in his company.

The car stopped in front of an antique shop. Wright got out, followed by Nigel and a plain-clothes sergeant, and rang the bell of a door at the side of the shop window.

"You didn't tell me you were bringing an army," Cyprian Gleed remarked when, after some delay, he had answered the bell. He was in silk pajamas and a new, cardinal-red dressing gown with a monogram on the breast pocket—an outfit which threw his scruffiness into strong relief but indicated why he was chronically short of money. He led them up a steep, nar-row staircase to the top floor; the second floor, he explained, was crammed with the overflow from the antique shop.

His sitting room was in a state of grotesque disorder, only equaled, in Nigel's experience, by that of Clare Massinger's studio. But, whereas her untidiness had something functional about it, or at least was the result of a single-minded con-centration upon essentials, the disorderliness here seemed almost pathological. An expensive radio was still playing a progressive-jazz record as they entered. Two golden hamsters scrabbled up the side of a basket chair. There were unwashed cups, plates and wine glasses everywhere; sheets of music on the floor; encyclopedia volumes on an elegant harpsichord in one corner; a dusty easel in another; a single ski and, rather oddly, a pair of boxing gloves hanging from a nail, in a third corner. An open door revealed the bedroom, an unmade bed with a woman's nightdress dangling from it, and a breakfast tray half concealed by a heap of clothes on the floor.

These fragments he has shored against his ruin, thought Nigel, feeling a little sorry for Millicent Miles' son. Inspector Wright had taken a chair on one side of the gas fire, which was protected by a high, old-fashioned nursery fireguard.

"I hope you will accept my sympathy, sir, in your bereavement," he formally but not perfunctorily began.

The young man's face twitched, then set into a contemptuous look. "We can dispense with the preliminaries. I don't like polite, meaningless words. My mother is no loss to me. She ruined my character—everyone will tell you that."

If it was intended to shock or otherwise impress the Inspector, Cyprian Gleed's speech fell remarkably flat. Wright, who had been gazing at him with the expression of a child seeing a new animal in the zoo, replied briskly, "Good. That'll save me a lot of trouble. You were here from 4:30 till 7 P.M. last Friday, I understand?"

"So I told your minion."

"And after that?"

"I couldn't wait about any longer. I was meeting some friends for dinner."

"You were expecting somebody who didn't turn up?"

"Yes. My mother."

Inspector Wright, who always contrived to look madly interested in the statements of those he was interviewing, had something to look interested about now.

"What time ought she to have arrived?"

"After she'd finished work, the idea was: between 5:30 and 6:30."

"It was by arrangement?"

"Ça se dit." Cyprian Gleed turned to the plain-clothes man, who had stopped dead in his shorthand at this point, and translated, "That is obvious."

"When had the meeting been fixed?"

"By telephone. The previous afternoon. I rang her at Wen-

ham & Geraldine, and asked her to look in on her way home
the next day."

"Any special reason for asking her, sir?"

Gleed's white teeth gleamed behind the scrubby beard. "Of
course there was. I never greatly cared to meet my mother
socially."

"What was the purpose of this, er, appointment?"

"Money. I wanted her to dish out."

"But she'd just refused you. That very morning," Nigel
put in.

"I fancied she would be more amenable later on."

"Why?"

"*Christ!* . . . I beg your pardon, sir." It was Wright's sergeant, upon whose notebook one of the hamsters had suddenly
materialized. "Lumme, what *is* this? Shoo! Get off."

"Don't get excited, Fenton. It's a golden hamster. Give it
one of your pencils to suck," suggested Inspector Wright.

But Fenton had already swept the animal off onto the floor,
and the next moment he was defending himself against
Cyprian Gleed, who had sprung at him, shouting, "You
bloody, clumsy oaf!" and was scratching at his face like a
woman. Nigel, who happened to be nearest, pinioned the
young man's arms and deposited him in the chair from which
he had leaped.

"Fenton," said Wright, winking at his subordinate, "no
roughhousing here, please. And you must not be unkind to
the animal creation, Hamsters are very pretty, inoffensive
little beasts. Have you had these long, sir?"

Scowling, Gleed answered, "My mother always forced me
to keep pets. They're supposed to be good for maladjusted
children. Substitute love-objects for the emotionally under-
nourished, if you see what I mean. Now I've got into the
habit."

"I see." Wright stroked the back of the other hamster.
"You were saying your mother would be more amenable later
on. Why did you think so?"

"Ask Strangeways. Perhaps he has a theory."

I wonder why he dislikes me so much, thought Nigel. He decided to take up the challenge. "You were going to apply pressure—going to say you saw her tampering with the proof copy of *Time to Fight?* Is that it?"

Cyprian sneered at him again. "'Apply pressure'! That's just like your generation. Tying pretty ribbons round ugly truths. Why not say 'blackmail'?"

"If you prefer it," replied Nigel, rather nettled. "My generation doesn't make a virtue of boorishness."

"And *was* your intention to blackmail your mother?" Wright asked.

"That's Strangeways' theory."

The Inspector let it go. "You didn't think of ringing her up on Friday evening, when she was late for her appointment?"

"She was working at the publishers', and their switchboard closes down at 5:30."

"And after that—Saturday and Sunday?" Wright's alert, friendly, terrierlike expression was unaltered. "Since you needed money so badly, I assume you tried to get in touch with her over the weekend."

Cyprian grinned. "Is this a pitfall that I see before me? I always need money badly. But it could wait—it wasn't all that urgent. Anyway, I was on a blind over the weekend. I started getting drunk at 7 P.M. on Friday. I remained drunk till Sunday morning. And from Sunday afternoon till Monday morning I was in bed with—or, as Strangeways would probably say, entertaining—a young lady."

"Miriam Sanders?" inquired Nigel.

"Correct."

"Do you benefit by your mother's will, sir, may I ask?"

"You may. I imagine so, but I don't know for certain. She detested me. On the other hand, she was a conventional-minded woman and would think that money should be left to one's flesh-and-blood." Cyprian Gleed screwed up one corner of his mouth. "And now, may *I* ask a question?"

"Certainly, sir."

"Have you, in the course of your career, come across any matricides?"

"Lord bless you, yes. Two or three. It's comparatively rare, though, I agree."

The young man seemed rather deflated by the Inspector's breezy reception of his query, and did not pursue the matter. Wright began taking names and addresses of people in whose company Gleed had spent the Friday night and Saturday. His sinus jabbing at him again, Nigel lay down on a settee and squirted nose drops up his left nostril, averting his eyes, after one glance, from a wash painting on the ceiling, which the artist had mercifully left unfinished, of a satyr at grips with a nymph. Perhaps I've underestimated young Gleed, he thought: he appears to have brains, and some courage—if his effrontery is a kind of courage and not just a product of living in a fantasy world. Sitting up, Nigel gazed round the fantastic room again, so deeply occupied with his own thoughts that, when Cyprian Gleed asked, "Well, do you like my flat?" he uttered without premeditation what was in his mind—

"It looks like a museum of false starts."

Cyprian's eyes rolled up and round, in that way reminiscent of his mother's mannerism; then they went dead, as he surveyed Nigel for a few seconds without speaking.

"Now, about the inquest, sir. You will be notified of it shortly. And the funeral arrangements—I daresay you would like your mother's executor to take them off your hands."

"Executor?" asked Cyprian dully. "What? Oh yes. For a moment I thought you'd said 'executioner.' "

Fenton, not for the first time, breathed heavily over his shorthand notes, his whole expression registering outrage and antipathy.

"Had your mother any enemies, sir?"

"I wondered when we were coming to that. Dozens, I

should think. If thoughts could kill, she must have borne a charmed life."

"But you've no reason to suspect anyone in particular? No threats, or—?"

"I tell you who had good cause to wring her neck. Basil Ryle. She was playing him up. Love me, love my books. God! But I shouldn't think he knew it. Besotted ass!" The venomous disgust in Gleed's voice made even Inspector Wright glance at him sharply. Oh dear, thought Nigel, this is Hamlet and his Mum all over again.

Further questions eliciting no more information about Ryle, or any other possible suspects, the three men left.

"Well," said Wright, when they were in the police car on the way to the dead woman's house in Chelsea, "what do we make of him?"

Fenton could contain himself no longer. "Downright immoral, I call it! Talking about his mother like that; and then what he was up to with his girl friend. Shameless little bastard!"

Nigel said, "I think he's rather pathetic."

"I think he's damned dangerous," Wright cut in. "I don't like engines rushing around without drivers. How clever is he, though? Clever enough to invent that story about blackmailing his mother?"

"Doesn't give him much of an alibi, sir," said Fenton.

"No. But a chap who expects to screw money out of his mother by threatening to expose her over that proof-copy affair—well, he wouldn't go cutting her throat till he was sure she wouldn't play. Is that how he wants us to reason?"

"I don't see him as a throat cutter," said Nigel. "He'd think up something more subtle and long distance."

"He acted violent enough just now," Fenton remarked.

"Yes, but that was off-the-cuff, my lad; and besides you'd been brutal to one of his dear little pets."

"Brutal! Well I—"

"Whereas the killing of Millicent Miles was a premeditated

affair," Wright continued. "Why 'pathetic,' Mr. Strange-ways?"

"Oh, that room! Full of things he's begun and never finished—pursuits he's taken up and dropped. Trying to prove to himself that he can do something really well, and invariably failing—do something except drink and cadge and seduce girls."

"'A museum of false starts,' eh?"

"I'm sorry for the children of successful people. They so often don't inherit their parents' talent, or stamina, or whatever it is that has made them successful. But they desperately want to impress their parents, somehow or other."

"By cutting their throats, for instance?" said Fenton.

"Fenton, don't be crude," Wright snapped. "Sarcasm doesn't become you."

"No, sir. Beg pardon."

"Were you a problem child, Fenton?"

"No, sir."

"Never threw harmless animals around the room when you were a boy?"

"No, sir," replied Fenton in a repressed tone. Inspector Wright was popular with his subordinates; but, as they said, "you had to watch him"; his reactions were not always predictable, nor could you always be sure if he was pulling your leg or tearing you off a strip.

Millicent Miles' house in Chelsea owed more to an interior decorator than to its late mistress. Of course, she had only been living here a few months; but the rooms held no personal flavor at all—they were like demonstration rooms in some exhibition of Gracious Living. If they gave any other positive impression, it was that impression of unshared life which one often receives from the houses of widows, ex-mistresses or businesswomen.

Nigel translated while Inspector Wright questioned Miss Miles' German maid. Hilda Langbaum, a stocky, flaxen-haired Fräulein, was at first frightened and over-obsequious, then—

since the English police showed no inclination to bully her—grumbling, voluble, and rather malicious about her late employer. Miss Miles was not sympathetic: she had been by turns brusque and inquisitive; she kept everything locked up; she was erratic in her comings and goings, inconsiderate about household arrangements; she was insufficiently appreciative of Hilda Langbaum's excellent cooking; she used bad language.

Yes, Mr. Gleed had come to the house occasionally, but not during the last few weeks. No, her mistress had seemed her usual self when she left for the office on Friday morning. Men visitors? Yes, sometimes alone, sometimes for small dinner parties: their names could be found in the engagement book, which was on Miss Miles' desk. The most recent visitor had been Mr. Ryle; he arrived about 9:30 on Thursday night; Hilda did not know how long he had stayed—she herself went up to bed soon after 10. Had a Mr. Protheroe been to the house? Not to Hilda's knowledge. Or any other members of the publishing firm? Mr. and Mrs. Geraldine and Miss Wenham were invited to dinner in September, but had not visited the house since.

At this point Mr. Deakin, the dead woman's solicitor, arrived, by appointment, to examine her papers with Inspector Wright. The two men repaired to the study for this purpose. A key had been found in Miss Miles' handbag, which Hilda Langbaum identified as the key to the desk drawer in which her mistress had kept all the other keys. The Yale to the side door of Wenham and Geraldine's was not to be found in this drawer, however, and Fenton was sent to search the other rooms for it. The course of the investigation, Nigel realized, must depend considerably for the present upon whether Millicent Miles or someone else had "borrowed" this key; if she had not done so herself, then it seemed probable that the murderer must have got hold of it; and why should he need it, except to enter the publishers' office after the main door had been locked at 5:30? The partners and Stephen Protheroe already possessed keys to the side door, which suggested that the murderer must be either some other member of the firm or an outsider conversant with its habits.

Nigel mooched after Fenton for a few minutes, then returned to the study. Here Inspector Wright handed him a desk diary, indicating two recent entries: "R., 9:30." was scribbled in last Thursday's space—that would be Basil Ryle's visit. Friday showed a luncheon engagement, and below it the laconic entry "Thorsday?!" Nigel pondered on the question mark, the exclamation mark, and the revealing pun, as he drifted down to the basement kitchen.

Hilda Langbaum was making herself a cup of coffee, and Nigel accepted one too. The girl was clearly worried about her immediate future; Nigel told her he would arrange with the solicitor that she should receive wages up to the end of the month and be allowed to stay on in the house for a week or two, if she so wished, while looking around for another job. Her gratitude for this small kindness was excessive. Nigel drew her on to talk for a while about her home in Nuremberg. Confidence being established, he led the conversation, via Hilda's fiancé whose photograph she showed him, to Basil Ryle. Hilda became at once sentimental and melodramatic. Such a good young man. So deeply in love. And the way she had treated him! A cold woman, a heartless, a mocker. A German lover would have beaten her, made her go down on her knees; but the Englishmen were soft; they did not understand that the man must be master; they ran away with their tails between their legs, like Mr. Ryle the other night.

Hilda had not wished to speak of this to the police, but Mr. Strangeways was different, sympathetic.

"When I went upstairs to bed," said Hilda, encouraged to continue, "I heard their voices in the drawing room. I do not understand your language, but I could not help listening. She was lashing him with her tongue. It was like a whip of ice. He spoke little: he sounded bewildered, shocked, as if—as if an angel had suddenly spat in his face. He was pleading then, I think; but she laughed—oh, very ugly her laughter was. And the next minute that poor Mr. Ryle hurried from the room and ran downstairs out of the house. He did not see me. He looked quite blind; and deathly white in the face, like a specter."

IX *INSERT*

That afternoon Nigel settled down with Millicent Miles' autobiography, which he had picked up at Scotland Yard when Inspector Wright had finished his work in the dead woman's house. The search there had brought no clues to light, unless the cryptic entry in her diary for Friday afternoon—"Thorsday?!"—could be called a clue. The solicitor having stated that he had drawn up no will for his client, and no will being found among her papers, there was a strong presumption that she had died intestate. Nigel hoped that the autobiography would be more revealing than the Chelsea house; indeed, he approached it with the liveliest interest. Millicent Miles had possessed one quality of the great writer— that negative capability which enables its possessor to take on the color of his environment, surrender himself to the personality of another. With her it was no more than a chameleon trick, and a trick which she had learned to exploit. But she had been one woman to the infatuated Basil Ryle, and a quite different woman to the sardonic Stephen Protheroe.

What was she to herself? Would the autobiography reveal the real woman behind this fluid personality, or would it offer only another illusion—a Millicent Miles cunningly adapted to the fantasies of her avid readers?

One thing was plain from the start: the reader would be taken into her confidence. She gave the effect of singling him out, drawing him aside, telling him things that were not for every ear. It was done with quite remarkable plausibility; and you had to keep your wits about you to perceive that this intimate address in fact disclosed far less than it appeared to. Her writing has the surface frankness of a mirror rather than the candor of transparent glass, thought Nigel. As he read on, he began to realize that every incident, every person in the book was a mirror set at just the right angle to reflect its author most becomingly. When she confessed some childish peccadillo, she conveyed the impression of a high-spirited girl, a jolly, romping tomboy whose escapades one could not but view with a smile of indulgence. When she hinted, ever so delicately, at dissension in the home, one could almost hear the rustle as a not quite opaque veil was drawn over her parents' less agreeable aspects; how sensitive the poor child must have been, you were led to think; and how wonderful she is to bear no resentment for the way she must have suffered.

It really is a masterpiece of calculated dishonesty, Nigel reflected; but before long he changed his mind, judging that Millicent Miles, like every best-seller of her type, believed implicitly in every word she wrote, at any rate while she wrote it, and was sublimely unconscious of contradictions or self-deception. That she could combine this unawareness with so skillful a manipulation of the reader's responses argued a most subtle sleight-of-mind—an egotism and fundamental irresponsibility truly formidable.

Nigel wondered how Miss Miles had proposed to explain to Basil Ryle the discrepancies between the story of her adolescence, as she had told it to him, and as she related it in her book. According to the book, though her father, a commission agent, had certainly gone bankrupt, he was neither a drunkard nor a sadist; she pictured him as a jovial, bumbling, fumbling character, whose sins were of omission rather than commission. Nor was her mother represented here as a slut and a domestic tyrant; she was, instead, a strait-laced woman of the Baptist persuasion who had done no worse than nag her husband incessantly and fail to understand her daughter's girlish aspirations. Nevertheless, though the treatment of these two in the early chapters was the reverse of satirical or denunciatory, an atmosphere of sweet forgiveness suspired from between the lines—a kind of holy hush which suggested that the author had had a good deal to hush up. Again, she had certainly got a job as shop assistant at the age of seventeen; but it was in a bookshop, and she did not run away from home to do so.

Nigel could hardly wait to get to Millicent Miles at nineteen, when—so she had told Basil Ryle—she was seduced and gave birth to a stillborn baby. With what convulsions of euphemism would she handle that passage of her life? Or, more likely, it had been another outrageous fiction invented to enlist the sympathies of poor Ryle. However, compelling himself not to skip, Nigel plowed on through the suburban tennis parties, the droll accounts of school or family friends, the young Millicent's lush responses to the Beauties of Nature and the Heritage of Great Literature, her parents' incapacity to "understand" her, the wincing sensibility which she hid from them beneath a hoyden's manner, the (carefully-edited) facts of budding womanhood, the clothes she had worn and the thoughts she had thought.

All this Nigel found sufficiently unreal. And the unreality was heightened by the author's habit of naming certain people with an initial capital and a dash. This secretiveness, tiresome in itself, became doubly so when Nigel discovered that the initials were fictive: in the margin, each time a new character appeared, the author had penciled an initial capital, different from the one which appeared in the text. The annotations, Nigel presumed, were an "aide-mémoire" and gave the correct initial.

Halfway through Chapter Four, he found a note from Inspector Wright pinned to the next page. "A capital G has been erased from the margin opposite line 19." A little flutter, like a cat's-paw on a calm sea, ran along Nigel's nerves. He was in contact with the murderer. The conviction was irrational—after all, Millicent Miles could have erased the G herself; but Nigel felt certain this was the murderer's work. Slowly he read the paragraphs in question, straining his inward ear to catch their overtones.

It was a few weeks after my eighteenth birthday that I met the man whom I shall call Rockingham. He came into the bookshop one afternoon, and we got into conversation. Little did I realize, as we discussed the latest batch of novels upon the shelves, serene and mysterious in their dainty jackets, that this shy, gangling young man would one day be a power in the world of letters. It was enough for me to discover, as I soon did—for there must have been already an affinity between us, which enabled me to overcome our mutual shyness and draw him out—that Rockingham was a fellow craftsman: he had even had work published in magazines. Ah, the magic of print to the young aspirant! I wonder does the modern generation, who have fearlessly destroyed so many obsolete shibboleths, understand what it has lost by discarding too the fine, youthful, eager virtue of hero worship? To me, then, a writer was a god. And if this particular god transpired later to have feet of clay, no shadow of premonition darkened my bliss at this first *rencontre*.

I sent him a short story I had written, and he returned it with criticism and encouragement that were worth more than a king's ransom to the diffident girl I then was. When I myself became "famous," and budding authors began to shower me with their prentice efforts, I remembered what Rockingham had once done for me. I resolved to set aside one precious hour every day to help these correspondents. Busy as my life has been, and with a full share of the anguish which falls to every woman's lot, I have always kept this resolution. My reward has been the gratitude of literally hundreds of aspirants to the laurels. This gratitude, no less than the loyalty of my faithful readers, effaces for me the sneers of the so-called highbrows. It makes me feel loved—and what woman does not want to feel loved? And it makes me humble.

It was not very rewarding, thought Nigel. All he could

extract from this chunk of stale, oversweetened self-praise was
that "Rockingham" had had feet of clay, that his real name
began with G, and that he became "a power in the world of
letters."

"G." It could stand for "Goggles," who might or might not
be Stephen Protheroe; it could hardly stand for Miss Miles'
second husband, the racing motorist. But again, bearing in
mind her practice of referring to gentlemen by their surnames
alone, it might stand for Arthur Geraldine.

Nigel read on, hoping to find some explanation of "Rockingham's" feet of clay. His name, however, did not recur in
the chapter. Halfway down the last page but one of this chapter, a new paragraph started:

So, as a writer, I was trying out my fledgling wings. But something else in me was budding too, ready to blossom at a touch of
the sun. I was nineteen, an ardent, inexperienced girl, my womanhood struggling to emerge as a butterfly from the chrysalis. What
can a mere man—even the most sensitive and upright man (and
there are some of them still to be found, thank God)—what can a
man know of the sweet bewilderment, the radiant expectation, the
quivering, throbbing ecstasy which possesses a girl's heart when
first it turns to love, as a heliotrope to the sun? Ah, magic days,
when we are in love with love! When everything is tinged with
"the light that never was on sea or land"! Looking back on that
time now, when the wound is healed and the bitterness has been
long forgotten, I can say with the poet, "So sad, so strange, the
days that are no more."

Yes, I fell in love. I brought to love all the pent-up ardor, the
desire to give and give and never count the cost, for which my life
at home had, alas, offered no outlet. The man I chose has appeared

Nigel turned the page. He did not need the note, which
Inspector Wright had pinned on the next sheet, to tell him
that this was the one the murderer had inserted. It was just
perceptibly whiter than the pages preceding and following it,
which had been stacked for months in the dusty room at
Wenham & Geraldine's. Unless Miss Miles had herself re-

moved the last page of this chapter recently, and retyped it, no one but the murderer could be responsible; and Miss Miles' method of composition, rapid and uncritical, rendered it extremely improbable that she should have redrafted just this one page of the typescript. Besides, there was the evidence of the sheet found in her typewriter after the murder, taken out and then replaced out of alignment, to strengthen the likelihood that the murderer had done some typing after committing his crime. Nigel read the last page of Chapter Four:

often to me in my dreams, since the day when the destiny which had brought us together tore us apart. But I never saw him again in the flesh. To meet him, even after all these years, would be unthinkable. Yet we seemed twin souls to each other! I loved him—yes, I confess it—as I have loved no man since. I gave him . . . everything. Freely, proudly, with all the passion of a nature starved for love, did I give. We were both poor, struggling. Marriage was out of the question. Perhaps—who knows?—if I had had a child by him, we might have outfaced the poverty, survived the disillusions that the flesh is heir to, and gone through life together hand in hand. But an inscrutable Providence decreed otherwise.

Toward the end of that year I fell seriously ill. Kind friends lent me the money to go abroad and spend some months in a sanatorium. When I returned, he, my beloved, was gone. I will draw a veil over my sufferings. I can be thankful for them now, since they gave me the deep sympathy, the understanding of human pain and human courage, which have enabled me to help others through my books. But at the time, with nothing of his to remember him by, I lived through an eternity of loneliness, emptiness—an ice age, a desert of stone. Even today, my pen falters as I trace these memories—

> "Deep as first love, and wild with all regret;
> Oh death in life, the days that are no more."

Nigel read on to the end of the typescript. It carried Miss Miles' life as far as her third husband. There was no reference to this husband's befriending of the young Basil Ryle, nor to the episode when the authoress had been hazed in General

Thoresby's mess. The birth of Cyprian Gleed was described with profuse clinical and emotional detail; thereafter, he served chiefly as a peg upon which to hang theories of child upbringing or lush sentiments about motherhood. The effect upon Nigel of some seventy thousand words of Millicent Miles was to make him long for the company of a real woman. He rang up Clare Massinger, and invited himself to dinner.

After they had eaten, he handed her the two concluding pages of Chapter IV. She read them with visibly growing distaste.

"What's all this about 'my pen falters as I trace these memories'?" was her first comment. "I should have thought she'd rattle the stuff off on a typewriter."

"So she did. It's author's license. You could hardly expect her to say 'my typewriter trembles.' Anyway, she didn't write that last page."

Nigel explained. Clare's voice went into a high pitch, as she said, "But it's horrible, isn't it? Do you mean to say he sat down at the typewriter and—while she lay weltering in her gore?"

"It seems likely. The join's pretty neat, don't you think?"

"The join?"

"Yes. The page she wrote ends 'The man I chose has appeared' . . . The next page begins 'often to me in my dreams.' Why did the murderer have to substitute this last page at all? Presumably because the one for which he substituted it gave us a clue to his identity, even his motive."

"Yes, that's obvious."

"Perhaps what she had written was 'The man I chose has appeared *already in these pages*.' That would narrow things down quite a bit. In fact, it would point directly at 'Rockingham.' "

"Who on earth is Rockingham?"

"Oh, of course, you need to read this." Nigel handed her the page from the middle of the chapter, and told her about the initial G. erased from the margin. Clare perused this page, then dropped it from between finger and thumb onto the divan beside her.

"I never knew people could write like this—seriously, I mean."

"My darling Clare, do you never read the shiny magazines?"

"Good gracious no. They're written by career girls for veneer girls, aren't they?"

"Now look at that last page again. Suppose the murderer didn't write it merely to conceal his identity. Suppose it's aimed positively to put us off a certain track. What track would it put us off?"

Clare reread the page, brooding over it like a witch, the black hair tumbling about her wrists.

"Well," she said at last, "there's this about their not having had a child. Yes, and it's sort of repeated lower down—'nothing of his to remember him by.' Is that it?"

"Clever girl. Basil Ryle told me that Millicent had told him she was seduced at the age of nineteen, deserted by the man, and had a stillborn child."

"I don't see how a stillborn child, nearly thirty years ago, would be a motive for murder."

"But if she was romancing about that part of it? If she had a child, and it lived?"

"Well, yes," said Clare slowly. "But in that case it's she who'd have a good reason to murder the chap, not vice versa."

"Not necessarily. If this chap turned up again recently in her life, if he was a man who couldn't afford a scandal, she might have been blackmailing him over their child. Blackmailers do get murdered."

"Ye-es, I see that. But could the darling of the lending libraries afford a scandal either? If I was the chap, I'd just have called her bluff. I say, you don't think Basil Ryle is her son?"

"My dear Clare, what a farcical notion!"

"You're only saying that because it hadn't occurred to you," replied Clare demurely.

"But she was leading him on, as a lover—actual or prospective."

"He discovers she is really his mum, and in a fit of revulsion murders her. Just like a Greek tragedy."

"Do be serious, Clare," protested Nigel, but a little uneasily. "Ryle knows all about his parents."

"It's a wise child who— Well, who's your candidate for Rockingham?"

Nigel got up and began stalking round the studio, picking up objects from shelves, staring at them unseeingly, and putting them down again in the wrong place—if any place, in Clare's habitat, could be called wrong for any object.

"Protheroe or Geraldine," he said at last. "It *ought* to be Protheroe—you see, I feel in my bones that the murder must be tied up with the libel affair."

"Nigel, sweet, you're being utterly incoherent."

"Assuming it was Stephen Protheroe who fiddled with that proof copy—"

"*Stephen*? But last time we discussed it, you said—"

"Never mind what I said last time," Nigel tetchily replied. "Can't I change my mind? Stephen had far the best opportunity to tamper with the proof. And then there's the evidence of Bates."

"Shall I leave the room so that you can talk to yourself in private?"

"Sorry, love. Bates was Production Manager in the firm. Stephen has no tolerance for bores, and Bates—according to him—was a Grade A bore. Yet when he's finished reading the proof, instead of getting a secretary to take it to Bates, Stephen goes along with it himself and has a jolly chat with Bates while it's being wrapped up to send to the printer. That seems to me highly suspicious."

"Why?"

"Why should he expose himself to five minutes of boredom unless he wanted to make sure that Bates didn't glance at the proof before dispatching it? The séance with Bates was the last step in an artful campaign to get the book published with its libelous passages all alive and kicking."

"All right, then," said Clare after a dubious pause. "So Protheroe is the murderer?"

"That's just the trouble. Stephen wouldn't care a damn if Millicent Miles threatened to tell the world she'd had a child by him. Arthur Geraldine, on the other hand, might. But one simply can't imagine Geraldine plunging his firm into a libel action."

Clare Massinger rose from the divan with one of the swift, sinuous movements that made her hair swirl like smoke on a gusty day, and drew Nigel down beside her.

"Poor boy. Your head's bad, isn't it?"

"Yes, this bloody sinus—"

"I mean, you're not cerebrating on all six cylinders."

"Oh. I thought you were giving me some womanly sympathy," Nigel complained.

Clare ignored the appeal. "Suppose Millicent Miles did have a living child by X. It'd be—what?—nearly thirty now. Suppose she threatened its father to tell it—call it 'him'— tell him that he was a bastard."

"Nobody minds being a bastard nowadays."

"Oh don't they just? In some circles it's thought a terrible disgrace."

Nigel went off at a tangent again. "I can't get it out of my head that Stephen must have known Miss Miles in the old days. The first time I met him, he said, 'She always—she hates puns.' Was it a near slip of the tongue? 'She always hated puns'—is that what he was about to say? Later that day he jeered at her, 'I thought your births were always painless.' Oh, I know the phrases needn't have meant more than they said on the surface. And Millicent told Ryle she'd never met Stephen till this summer. But I got this strong impression of a familiarity between them. If it existed, why were they at such pains to deny it? Well, I can see only one answer to that."

"What answer?"

"That they collaborated over tampering with the proof copy, that their motive for doing so sprang from something in their past association, and therefore the association must be kept dark."

"It's all madly theoretical, isn't it?" said Clare. "I'll fix up for you to meet Mrs. Blayne—she's the woman who knew Miss Miles in her youth."

"Thank you. That might help. But then there's old shark-face—Geraldine," continued Nigel abstractedly. "Millicent told me she'd met him years ago, 'under rather strange circumstances.' And why did he let her go on occupying that room at the office and making a general nuisance of herself? And he did rather firmly warn me off digging up the past. He's married—a far better subject for blackmail than Stephen, I'd have thought, if blackmail was what Millicent got murdered for." Nigel, frowning, picked up one of Clare's little clay horses and rubbed it against his cheek. "It's odd, the way people will commit murder to achieve security. As though one could ever buy peace of mind with someone else's blood."

"Most people don't see beyond the next step. Particularly when the next step is a precipice they're pushing somebody over."

"If, on the other hand, it was Stephen Protheroe alone who fiddled with that proof, and if Miss Miles could blackmail him about it—"

"But why should she?"

"By telling him he must withdraw his opposition to the firm's reprinting her books, or she'd expose him. Awkward dilemma. He's a man of integrity—where books are concerned, anyway. But his security would be at stake. Liz Wenham told me his whole life was bound up with the firm. They'd have to sack him; he'd never get another job in publishing; and he's dried up years ago as a writer. Oh yes, if security was the motive for this murder, Stephen Protheroe's the obvious suspect."

X REVISE

It was a tradition of Wenham & Geraldine, handed down from the firm's earliest days, that on the last Wednesday of every month the partners and the reader should dine together. In more expansive times a private room at Skimpole's used to be engaged for these occasions; but a bomb had destroyed Skimpole's in 1944, and it being unthinkable that the firm should patronize any other restaurant for the time-hallowed observance, these dinners had been held since then in Arthur Geraldine's flat. Normally they were confined to the firm's ruling circle. But twice a year, the legendary James Wenham had laid it down, guests might be invited—a fellow publisher, a distinguished editor, a literary-minded bishop perhaps, even an author. When one of the firm's authors received such an invitation, it was understood to be a notable honorific, a medal for long service and good conduct, particularly the latter.

The strength of this tradition may be gauged by the fact that, in spite of a murder on the premises, neither to Liz Wenham nor to Arthur Geraldine did it occur for a moment that the monthly dinner should be canceled. Where the blitzes had failed, it was unlikely that poor Millicent Miles would succeed. The partners did, however, pay tribute to the out-of-jointness of the times by inviting Nigel Strangeways, although tonight's was not one of the twice-yearly guest dinners.

Arthur Geraldine had given Nigel the invitation this morning, in the manner of one who both asks and bestows a bless-ing. Though reporters, the police and his lawyer had been making for a harassed life, the senior partner had not lost his urbanity. Like many Irishmen, he was a natural actor, with a very hard core of realism, no doubt, beneath the smooth, fluent exterior. "Black tie, if you will," he had murmured. It was evidently part of a tradition: one must dress for dinner to keep up one's morale and self-respect in a jungle infested by wild authors, prowling literary agents, treacherous fellow publishers, and pigmy reviewers with their lethal blowpipes.

Dressing on Wednesday evening, Nigel reflected upon a day which had been most unsatisfactory. First, there was the interview with Jean, the girl for whom the lively Susan had been substituting in the Reference Library. Jean could add little to her friend's secondhand account of the altercation between Stephen Protheroe and Miss Miles; she was pretty sure that "Goggles" had been used in the vocative case; also, she had heard one shocking remark made by Miss Miles—"You're just impotent!" It was odd, the way the case kept reverting to this topic—did she have a child at nineteen? Did it live? Whose was it? But Miss Miles' vindictive remark prob-ably referred to Stephen's impotence as a writer: Nigel had heard her say to him, on another occasion, "I do *write*, any-way." Just before he left her, Jean said,

"Susan's got something on her mind. She hasn't told me, but I can sée it. I wish I knew—"

At this point her telephone rang. Mr. Ryle wanted her to look out two books and bring them down to his room. The interview was ended. Whatever it was that weighed upon the feather pate of Susan, Nigel had more important matters to deal with first—or so, regrettably, he assumed.

Stephen Protheroe was one of them. He looked up from a manuscript, with the dazed expression of a bookworm that has chewed its way through another inch of paper, as Nigel entered.

"What does 'Rockingham' mean to you?" asked Nigel without preamble.

"China. Also an English Prime Minister of the later eighteenth century. Why?"

"You didn't read Miss Miles' autobiography, then?"

"My dear fellow! I don't believe in going to meet horror halfway."

This shock tactic having failed, Nigel tried another. Yes, Stephen thought he remembered the particular row Jean had overheard. It had been about his opposition to the firm's reprinting some of Millicent's novels. Yes, she had started calling him "Goggles"; nicknaming was a feature of her lamentable schoolgirlishness; had she not called General Thoresby "Thor"? She was a woman addicted to overfamiliarity.

"We found an entry in her desk diary at home," said Nigel. "For last Friday. Very cryptic. The word 'Thorsday,' with a question mark and an exclamation mark after it. What do you suppose that could mean?"

Stephen's mouth made one of its nibbling movements.

"An appointment with the General? But she wasn't sure if he'd keep it?"

"General Thoresby has told the police there was no such appointment. He was out of London on Friday."

"Well then!" Stephen Protheroe threw up his hands.

"My own guess," said Nigel, "is that the entry referred to the General's book. A meeting with someone else about it. A showdown, perhaps. I'm convinced the murder and *Time to Fight* are linked somehow. The police don't seem very keen on the theory," he mendaciously added; "but those are the lines I shall work on. A missing link. Buried deep in the past, maybe."

If there was an opening here, Protheroe did not take it; nor had Nigel any desire, at present, to force his hand—or, for that matter, anything to force it with. Stephen was an elusive personality, warm and stimulating at times, but capable of switching off his attention in a disconcerting way. Now, like a fish darting into an aquarium grotto, he retired again into the manuscript he had been reading.

The rest of the day was no more rewarding. Nigel found Basil Ryle too busy, or unwilling, to see him. He finally tracked down, after several telephone calls, the Mrs. Blayne whom Clare had suggested as an informant about Millicent Miles' early days; but Mrs. Blayne, a J.P. and an indefatigable committee woman, could not make time to see him till to-morrow. Late in the afternoon, he had a talk with Inspector Wright at Scotland Yard. Wright's team had extended their search, over the last twenty-four hours, from Wenham & Geraldine's offices to the private residences of the partners, Stephen Protheroe and Cyprian Gleed; none of these raised any objection; and no bloodstained clothing or galoshes had been found, no sign of anything having been burned. The murderer, in any case, had had the whole weekend to dispose of the traces of his crime. In so far as the alibis of the five for Friday evening could be checked, they had stood up to the rigorous examination; Wright's men would now have to check their movements over the whole weekend; and of course there was the usual notification of laundries, dry cleaners, old-clothes shops. But at the moment it seemed as if the murderer had walked out of his bloodbath as clean as a needle. Arthur Geraldine's razors—he was the only one who used cutthroats—had been tested in the laboratory and found guiltless. The autopsy upon the dead woman showed that the time of death must have been between 4 P.M. and midnight on Friday, probably earlier rather than later—a finding which was consistent, Wright judged, with the nailing-up of the sliding window. Little had been proved, except that the killer was an audacious and a resolute individual.

Shortly after eight o'clock Nigel rang the side-door bell at Wenham & Geraldine's. Arthur Geraldine himself opened it, and took him up to the top floor in the lift. Nigel was introduced to Mrs. Geraldine, a slim, tall, well-preserved woman, with an excessively gracious manner, some ten years younger than her husband. Liz Wenham, Ryle and Protheroe were already there, conveying that atmosphere of intimacy blended with a certain disorientation and disengagement, which is so often generated when people who work together in a unit meet socially. Nigel, however, found himself fascinated less by the atmosphere than by the environment. The drawing room was a showpiece, a museum. On the mantelpiece, on shelves along three walls, in cleverly-lit corner cupboards, there stood a ravishing array of china. Nigel was no expert on ceramics, but he could recognize the quality of what he saw here.

"I'd no idea you were a collector. What exquisite things you've got."

Arthur Geraldine's eye lit up, the thin shark mouth softened. "I'm glad you like them. Yes, I suppose I'm a bit of a connoisseur"—he pronounced it, in the Irish way, "connossoor."

"Must be a life's work keeping them dusted," said Liz Wenham gruffly to her hostess.

"Oh, my husband never allows anyone else to touch them. Arthur, do give Mr. Strangeways a drink."

Geraldine was already displaying a piece to Nigel, his fingers curling round it, caressing it, as though the porcelain communicated to them its own delicacy.

"Of course. I'm sorry. Get me onto these things and I'd bore you for hours."

Basil Ryle was staring at the mantelpiece with a kind of smoldering animosity: one of those pieces, he might have been thinking, would have saved my Dad in the Slump. The face under his red hair was white and peaky, with dark rings under the eyes. He looked as if he was still suffering from shock: the shock of Thursday night, wondered Nigel, or of Monday morning? Or was it he who, on Friday evening, had used the razor? Strange to think a murderer might be sitting here amid these colored tiers of porcelain that glowed like herbaceous borders.

At the round table in the dining room he sat between Mrs. Geraldine and Liz Wenham. On the wall facing him hung oil paintings of two inscrutably bearded personages—the founder of the firm, James Wenham, and John Geraldine, his first partner. With a pleasant mixture of ceremony and informality, Arthur raised his glass and said, "the firm and its founder." Liz Wenham murmured the words devoutly as she drank the toast; it meant as much to her as, quite clearly, it meant all-my-eye to Basil Ryle, who made but a perfunctory response.

Nigel glanced at the side plates, noting their delicate apricot coloring.

"You have your treasures for use as well as for ornament, Mrs. Geraldine."

"Yes. But we only bring out the Rockingham set on these special occasions."

Nigel hoped he did not look like a man who has been struck a violent blow in the solar plexus.

"Is it a recent acquisition?" he asked, aware of Stephen Protheroe's quizzical eye upon him.

"Oh no. My husband specialized in Rockingham ware very early—before we were married." She gave an affected little tinkle of laughter. "It was quite like marrying into a china shop."

"Yes, I picked them up cheap in the twenties," said Geraldine. "You could get bargains still in those days."

"People didn't always know the value of their own possessions," put in Stephen—a remark which sent a little waft of uneasiness round the table. It seemed to Nigel a suitable moment for moving in.

"I glanced at Miss Miles' autobiography yesterday," he said.

Arthur Geraldine broke the somewhat dismayed silence with, "And what opinion did you form of it?"

"As literature, worthless. As an unconscious character study, fascinating."

"'Unconscious'?" said Stephen. "I'd have thought her whole life was a series of posed self-portraits."

"What is fascinating is to compare the self she exposed with the self she thought she was exhibiting."

"This is all too subtle for me," Liz Wenham said.

"And must we talk about the poor wretched woman just now?" said Mrs. Geraldine.

"Why not?" Basil Ryle muttered. "It's what we're all thinking about."

There was another awkward silence. Mrs. Geraldine had the rattled look of a veteran hostess who for once has lost control of her party. Ryle broke out again, "Why are we trying to pretend nothing has happened?"

"Because we're so damned civilized," said Stephen with a sympathetic chuckle.

"Where I come from, we called a spade a spade. And we didn't look upon policemen as guardian angels." Basil Ryle had already been drinking: his words, his glances round the table, were slurred. "They were at me last night. Why did you visit the deceased on Thursday? From information received, we understand you had a violent disagreement with her. What was the subject of this disagreement? Did you have another meeting with her, subsequent to the scene at her house? Did I, in other words, walk into her room here and cut her throat? Sweet Christ!"

A meaningless smile was frozen onto Mrs. Geraldine's face. Liz Wenham, her rosy-apple cheeks and bright clear eyes, together with her elaborate, unbecoming toilet of coffee-colored lace, making her look like a child in fancy dress, came to the rescue.

"My dear Basil, of course you didn't. That's quite out of the question. But you must pull yourself together. The trouble with you is that you will bottle everything up—it's just asking for an explosion." Liz Wenham's tone was pure and wholesome as a North Country beck. "You had a row with Miss Miles. All right. I don't blame you. But why turn the whole thing into a Strindberg melodrama? It's morbid. Worse than that, it's maudlin self-pity."

"Well, I must say—" Basil laughed shakily, subdued by this cold douche. "I expect you're right, Liz, but—"

"Of course I'm right. And I want to enjoy this delicious *tournedos*. If we must have a post-mortem, let's keep it for after dinner."

At this point Nigel led Geraldine to talk about the history of the firm, which the senior partner was glad to do, showing himself also a ripe anecdotalist. The dynasty of Wenham & Geraldine had been unbroken for over a hundred years. Arthur and Liz were the third generation. Liz went into the firm from Girton. Arthur succeeded a cousin, who had died in 1925; he had not intended to make publishing his profession, but as the nearest in line of succession to John Geraldine, after the cousin's death, he had thought it wrong to refuse. The strength of this dynastic feeling was evident in the way both Arthur and Liz talked about the firm; it could be called fanaticism, were it not so absolutely calm and self-assured. Personal ambition, if it existed in them at all, came second to the firm's prestige. No wonder, thought Nigel, the libel business was such a blow to them; but how far would either of them carry this loyalty? as far as, in the last resort, destroying someone who was a threat to the firm's good name? No, that's surely unthinkable. But if Arthur Geraldine is "G"—is "Rockingham"? If he had an illegitimate child and deserted its mother? Would not such a revelation, even nowadays, cause a respectable pillar of publishing to totter?

After dinner, when they had drunk their coffee in the drawing room, Mrs. Geraldine left them. It was the normal practice on these occasions, though tonight it was not done so that the partners could talk shop. As the door closed behind his hostess, Nigel felt a tension in the air: the other three were looking at him, expectantly and uneasily. He answered the unspoken question:

"Afraid there's nothing much to tell you yet. I had a talk

with Inspector Wright this afternoon; but he's a bit at sea—too few clues and too many suspects."

"Too many?" said Arthur Geraldine. "You mean we're all under suspicion?"

"I'm glad to hear it. I thought I had been cast as the murderer," said Basil Ryle.

"Really now, Basil! Just because you had a difference of opinion with—"

"Difference of opinion!" The young man laughed harshly. "That's like Dr. Johnson calling Ben Nevis a not inconsiderable protuberance."

Nigel eyed Ryle noncommittally. "It could be argued that your quarrel with Millicent Miles points to your innocence."

"A pretty paradox," said Stephen. "Pray enlarge upon it."

"The murder was premeditated and very carefully worked out. Ryle would hardly have time for that between Thursday night and Friday evening. But, if he'd planned it well in advance, he'd not call attention to himself by squabbling with his victim the night before. By the way, what were you squabbling about?"

"I prefer not to say."

"Oh, Basil, don't be so noble!" said Liz Wenham strenuously. "We're your friends, I hope. Even if you had to keep it from the police—"

"Very well, Liz, if you must have it," Ryle blurted out. "I was fond of her. I'd thought she—well, had some feeling for me. I found I was wrong. She disabused me of the notion very thoroughly."

"Because she'd discovered you couldn't be of use to her?" asked Nigel. "Over reissuing her novels? Something like that?"

Ryle nodded, miserably, his eyes cast down. Liz Wenham broke out: "And about time you were disabused, Basil. Trailing after that woman as though she was an angel of light!"

"You don't stop loving somebody because she turns out—turns out—" Basil Ryle's voice, almost inaudible, guttered down to nothing. No one could look at him. Stephen Protheroe broke the embarrassed silence:

"Well, Basil has been cleared. What about the other candidates?"

"Nobody has been *cleared*," said Nigel. "Not yet. The other candidates, as you call them, are in this room. Plus Cyprian Gleed, of course."

Arthur Geraldine looked shocked. "My dear fellow, are you suggesting that Miss Wenham, or—or I—?"

"The police are bound to suspect anyone whose alibi is unconfirmed. They go for the obvious first, and they are usually right. Cyprian Gleed is the obvious suspect, because he had a strong motive and his alibi is unsupported. But they can't do anything about it till they find the weapon, or some bloodstained clothing, or the spare key to the side door here."

"Ah, so that's why they've started asking about our movements on Saturday and Sunday," said Geraldine.

"Yes. The murderer might not have disposed of the weapon, etc., till then." Nigel paused for a moment. "You see, none of you has an adequate alibi even for the period when the murder was being committed. Mr. Geraldine was alone in his room from 5:20 to 5:50—these times are all approximate of course. Mr. Ryle was alone in *his* room from 5:25 to 6:00. Miss Wenham was alone in her room from 5:20 to 5:30; she then left the office—that's corroborated; but theoretically she could have let herself in again at the side door, killed Miss Miles, and still got to her cocktail party in Chelsea by 6 P.M. or soon after."

"But it's absolutely fantastic to suppose that Liz—" protested Geraldine.

"I'm simply telling you what's in Wright's mind. It has also occurred to him that a woman might wear galoshes, several sizes too big for her, to confuse the investigation. Well then, there's Protheroe. Miss Sanders confirms that he left the building at 5:20. We know that he caught the 6:05 at Waterloo, because his friends met the train at the other end. He says he walked to Waterloo over Hungerford Bridge. But he too could have popped back into this building by the side door, and—"

"No, he couldn't," Liz interrupted. "It's bolted inside till 5:30."

"I could have undone the bolts, though," offered Stephen, "*then* gone out through the reception room, then—"

"Stephen, I wish you wouldn't talk like that," Liz Wenham said, half scoldingly. Her affection for the little man had never been so evident to Nigel.

"All this popping out and popping in is nonsense," Arthur Geraldine declared. "You'd be spotted returning to the office. Particularly at 5:30, when a lot of the staff are leaving by the side door."

"I think it's morbid talking like this at all." Liz Wenham's face was flushed. "It's preposterous to suppose that one of us could commit a murder—that sort of murder, anyway."

"All right," said Stephen Protheroe. "We four are much too decent to kill even a Millicent Miles. So, by elimination, Cyprian Gleed must be the culprit. No doubt he has a water-tight alibi which the eminent sleuth, Strangeways, will finally puncture."

"No. He was alone in his flat from 4:30 to 7, he says. He can't prove this, and we can't disprove it—so far. He says he was expecting a visit from his mother, fixed up by telephoning her here the previous afternoon; he expected her sometime between 5:30 and 6:30; he was alone during the period because he wanted to have a private conversation with her."

"You think he made all this up?" asked Geraldine.

"If an alibi was in his mind, I'd have thought he could have cooked up something a bit better. And Miriam Sanders says a call did come through from him to his mother on Thursday afternoon."

"Well, that settles it then."

"I'm sorry, it does not. Cyprian could have rung up to make

an appointment with her here. Or he may have persuaded
Miriam Sanders to invent the telephone call."

"But—"

"She's his mistress. And, I fear, quite under his thumb."

"Miriam?" exclaimed Liz Wenham. "But she's a First in
History."

Even Basil Ryle joined in the laughter. Flushing, Liz added,
"I don't mean that history scholars don't go to bed with
people. But she's a clever girl, ambitious too—she'd never
take up with a worthless little runt like Gleed."

"Oh, Liz," said Ryle. "I'm clever and ambitious, and look
who I took up with."

"That's entirely different."

"It occurs to me," said Geraldine slowly, "if the telephone
call did come through, and if someone overheard it—or Miss
Miles could have mentioned to someone here that she was
going to have an interview with her son the next day—well,
it'd give the murderer—but of course that's pure speculation.
I—"

"What on earth are you dithering about, Arthur?" Liz
sounded pettish.

Stephen Protheroe grinned at her. "I'm perpetually being
reminded that Millicent occupied the adjoining room. As if I
needed a reminder! What Arthur was trying to say is that I
could have overheard the telephone conversation, known from
it that Gleed would be alone in his flat on Friday evening, and
arranged the crime for that period so as to throw suspicion on
the wretched youth."

"Oh really now, Stephen!" Geraldine began to protest.

"And did you?" asked Nigel.

"Arrange the murder?"

"Overhear the telephone conversation?"

Stephen Protheroe hesitated a moment before replying, "I
seem to remember her telephone bell going that afternoon: it
didn't ring often. But I couldn't have heard what she said."

At this point Nigel's sinus began to ache fiercely again, and with apologies to his host he disposed himself full length on a sofa to inject the nose drops, sympathetically watched by the others. Their faces, and the beautiful ranks of porcelain, swung in a semicircle, then swung back, as, dizzy for a moment, he resumed an upright posture. He found himself gazing at Liz Wenham's eyes, noting in them an abstracted, apprehensive look which he had never seen there before. Becoming aware of his scrutiny, she said,

"You ought to have it operated on. No good tinkering about with patent remedies."

"I've been advised against what they call surgical interference."

Arthur Geraldine leaned forward. "One thing I can't understand—why do the police think it happened between five and six?"

"Between 5:20 and 6, you mean. Protheroe didn't leave his room till 5:20." Nigel explained about the nailing-up of the sliding window. "That suggests the crime was committed while there were still people in the office: but your staff all leave not later than 5:30 on Fridays. The only danger, from the murderer's point of view, was that one of the partners might be working till later. In fact, two of you were."

Liz said, "I can't think why it has to be assumed that the murderer is one of us, or one of the staff."

"It's only assumed that the murderer was someone the victim knew, and was probably expecting—someone, at any rate, familiar with the layout and working hours of the office."

Basil Ryle had been looking more and more puzzled. "But do you mean to say—? However well she knew the murderer, surely she'd think it very queer if he started nailing up that window?"

"Oh, that was done *after* the murder."

"*What?* I thought you said it was done as a precaution against his being caught—"

"Not caught in the act of murder. Caught doing what he had to do after it."

The partners looked at one another in bewilderment.

"After the murder?" said Stephen Protheroe. "But that's fantastic! He wouldn't hang around—"

"Destroying his traces, you mean? Laying false clues, or something?" asked Geraldine.

Nigel was thinking fast. He had to make a snap decision. If the murderer was one of these four, he might be put on his guard or he might be rattled by what Nigel could say at this point.

"I don't know if—it's just a conjecture of mine," he slowly began. "Well, say a hunch. Suppose the secret of this crime lies in Millicent Miles' past. Suppose her autobiography contained a vital clue to motive. The murderer might have known this, or guessed it. If so, he'd want to find the page in her book that would give him away, and remove it. So he'd nail up the window—"

Liz Wenham almost exploded with impatience. "Are you seriously suggesting that he sat down, after killing her, to read through two hundred pages of typescript? He'd take the whole lot away, surely, and destroy it?"

"Not necessarily. A more subtle mind—and I think we're dealing with a subtle mind—might prefer to take away only the incriminating page or pages—"

"Well, are any missing?" asked Ryle.

"No. But he might have nailed up that window to give him time for typing substitute pages. One of your staff did hear someone typing in Miss Miles' room at 5:30 last Friday."

Arthur Geraldine exclaimed, "Oh, but that's too bizarre, my dear fellow. Surely you're wasting your time—"

"I don't agree at all, Arthur," said Stephen Protheroe. "I'm almost beginning to get interested in the crime myself. If Strangeways is right, it presents a pretty exercise in textual criticism: which pages, if any, can be shown on internal evidence to have been written by another hand? Or perhaps—" he smiled engagingly at Nigel—"you've already discovered that?"

"No. As I told you, I've only had time just to glance at the book. I'll have to get down to it tomorrow—no, I've got interviews then—the day after tomorrow."

Liz Wenham said, "But isn't it a job for the police?"

"Well, yes. However, Inspector Wright knows me pretty well; and literary detection isn't his line."

"I wouldn't mind having a go at this," declared Stephen with animation. "I'll have to read the darned thing sometime, anyway, if we're still going to publish it. Perhaps we could collaborate over the problem."

"Don't be so ghoulish, Stephen," said Liz.

"It might be useful," Nigel said, politely but noncommittally. Well, he thought, some sort of a trap has been baited. Now let's wait and see if anyone falls into it within the next twenty-four hours or so.

He was on the point of rising to go when Stephen Protheroe remarked, "Oh, by the way, what was that you were saying about Rockingham this morning? Something to do with the autobiography, I gathered."

"Yes. I happened to notice the name in an early chapter. Miss Miles gives some of the people in her book pseudonyms. She seems to have known this 'Rockingham' pretty well. I just thought, if I could find out who he was, I might get a line on her young days from him—you never know what may come in useful."

Arthur Geraldine's thin lips were stretched so tightly, Nigel observed, that they almost disappeared into his face.

XI BRING BACK ·

To Nigel, whose type figure of the public woman had been Dr. Edith Summerskill, Mrs. Blayne came as something of a surprise. Her voice was soft and melodious, her presence unformidable; she resembled some neat, shy, energetic bird—a dipper, perhaps—as she sat at the restaurant table, her black suit edged with white piqué at collar and cuffs, her head bobbing over the menu. It was the day after the dinner party, and Nigel had invited Clare Massinger to lunch as well.

When they had chosen their food—a lengthy undertaking with Clare, who was always thrown into a paralysis of indecision by a well-stocked menu—Nigel told Mrs. Blayne a little about the case in which he was involved.

"So, you see," he ended, "it's a matter of getting a line back into her early days. Clare told me you'd known her then."

"Oh yes, we were at school together. Wimblesham High School. I knew her quite well as a girl. Then I got a scholarship to Somerville, and we rather drifted apart. But I still used to see her from time to time in the vacations."

"When she was working in the bookshop?"

"Yes." Mrs. Blayne's neat head bobbed at him, and she came out with a charming little cascade of laughter. "Funny, the things one remembers. Have you ever seen an assistant in a bookshop actually reading?"

"No, I can't say I have."

"Well, she used to. Whenever I went in, she had her head in a book. Always a novel. I was rather priggish about the trash she read. Poor Millie! And I think she was jealous of me being at the university. It was an uneasy sort of friendship."

"Women's friendships always are," remarked Clare. "We're perpetually trying to improve one another, and resenting one another for it."

"But she had men friends too?" Nigel asked.

"Dozens, if you could believe her. But she was a wonderful romancer, even at school; she at least made you suspend your disbelief. The English mistress said once that Millie's flair for fiction was worthy of a better cause. Which could be her epitaph, I suppose, poor woman." Mrs. Blayne paused to sip her martini. "On the Bench, you know, I get girls of that sort up before me every week. For petty theft, shoplifting, and so on. They simply can't distinguish between reality and their own fantasies. And they'll always put the blame for their misdemeanors somewhere else—on their own bad luck, on parents who don't 'understand' them, on society which won't give them a break. I'm talking too much."

"Not a bit. It's helping to fill in the picture. You'd say Millicent Miles, as a girl, was a potential delinquent?"

"I hate that word. But, yes, I suppose she was. The difference was that she had an uncanny knack for staying out of trouble, for evading the results of actions for which she had been responsible."

"You mean, she threw the blame on other people?" asked Clare.

"It isn't as simple as that. She was a strong personality, of course, full of vitality and magnetism, what used to be called a tomboy. At school she was always the center of a group. But there was this other side to her—a streak of slyness—no, of incorrigible self-deception, so that, when she did something wrong, she could dissociate herself from it—pretend to herself it had never happened, and take herself in so successfully that sometimes other people were taken in too."

Mrs. Blayne paused, while they were helped to the first course. Then the reminiscent look returned to her eyes.

"Her father was a totally impossible creature. I suppose she caught it from him. Always living beyond his means—I remember my parents talking about it when he went bankrupt. Oh yes, that reminds me, it was just about then—the bankruptcy time—when I got a curious insight into Millie. It must have been our last year at school. She confided in me one day that a man had made a pass at her; we didn't call it that then, but Millie spared no detail. I heard later that she had confided it to quite a few other girls as well, and the story grew more lurid by repetition. I was a senior prefect then, and full of public spirit—a horrible little prig, in fact. I told her she must denounce this man to her parents. She said she did not dare—he'd threatened terrible things if she tried to expose him. So I said I'd report it to the head mistress. Millie begged me not to, but off I trotted. Fortunately, the head mistress was a sensible woman. She sent for Millie at once and confronted her with my evidence. And Millie got away with it, you know. She said she'd done it to pull my leg—it was all a joke—it just showed what an unhealthy mind I had, that I could take the absurd story seriously. But she had such an open, wholesome, antiseptic sort of air when she confessed to the leg pull, it really made me feel a complete fool—as if I *had* a foul imagination. She was reprimanded, of course; but the H.M. clearly accepted it all as a bit of not-so-clean fun on Millie's part, and of overzealousness on mine."

Clare asked, "Had she in fact made it all up?"

"I'll never know. She was quite capable of it: anything to dramatize herself. On the other hand"—Mrs. Blayne gave a little grimace—"she was a physically mature girl at that age."

"Did she tell you the man's name?" said Nigel.

"No. She said he'd come to the house when her parents were out. He was a commercial traveler—no, I remember now—a representative of the *Daily Sun*; you know the circulation scheme newspapers ran in the twenties: take out a year's subscription to the *Daily Sun*, and we give you absolutely free a set of Charles Dickens' novels."

"She didn't describe his appearance?"

"I don't think so. Just one thing stuck in my mind, though—Millie saying to me, with bated breath and a quite devastating sexual precocity, 'Julia, never trust a man who has thin lips.' "

"Indeed?" said Nigel, poker-faced. "Well, well. It's a small world. But not so small that there aren't thousands of thin-lipped men in it."

"You mustn't mind him, Julia. He has a bad habit of talking aloud to himself. And don't let your food get cold."

As Mrs. Blayne caught up with her eating, Nigel pondered upon one of the entries in *Who's Who*, which he had consulted with the forward Susan leaning against his shoulder. Arthur Geraldine's entry gave no information for the period between "Educated at . . ." and "Joined staff of Wenham & Geraldine, 1925."

"What year did this happen—the *Daily Sun* representative making a pass at Millicent Miles?" he presently asked.

"It was our last year at school: the summer term of 1924."

"Yes, it would fit."

"Mysterious old cuss, aren't you?" said Clare affectionately. "Did you go to the Miles house much?"

"Not a great deal. Millie was a bit ashamed of her parents, I think. Too coarse-grained for the sensitive soul she made herself out to be—oh dear, that's rather uncharitable. It was the usual hideous little semidetached. Mrs. Miles was terribly houseproud, though. The scouring and dusting type. And oh so respectable. One took tea in the front parlor, I remember—cold as a morgue, and all cluttered up with atrocious objects. About half the room was taken up by a huge glass-fronted cabinet filled with plates, which I've no doubt were never used."

"Would they have been, by any chance," inquired Nigel, gazing noncommittally down his nose, "plates of a delicate apricot color?"

Mrs. Blayne looked prodigiously startled. "Yes, they were. But how on earth—?"

"I should have warned you," said Clare. "Nigel has strange powers. He is a species of warlock."

"You were visualizing the plates, weren't you, Mrs. Blayne? Extrasensory perception is much less rare than one supposes," Nigel equivocally remarked. "They were sold up, presumably, when Mr. Miles went bankrupt."

"That I can't tell you. I never actually visited the house after that scene with the head-mistress. I felt pretty sore with Millie over it, and chucked her for a bit."

"But later you came together again."

"Yes, I was sorry for her about the bankruptcy. Not that I really needed to be—she made an enthralling drama out of it for herself."

"How long did she work in the bookshop?"

"About two years, I think. Then she got ill—yes, that must have been in 1926, my second year at Oxford. She told me she'd got T.B. and was going to a sanatorium—in Switzerland, was it?—I'm very vague about it all now. Soon after that, my own parents moved away from Wimblesham and I lost touch with her completely."

"What about men friends during the bookshop period?"

Julia Blayne did not think she could help much here. Millicent, she suspected, would have been as secretive about any real affair as she was communicative about imaginary ones. Besides, in the circles where Millicent moved at that time, bankruptcy was a disgrace, and the local youth fought shy of her for a while—Mrs. Blayne remembered her friend's resentment over this.

"Did she talk to you about her literary ambitions at all?"

"Oh yes. In fact she inflicted quite a number of her compositions on me," Mrs. Blayne dryly answered.

"She never mentioned striking up a friendship with a young writer, a man who helped her with her writing?"

"No. But, as I say, she could be very secretive."

Clare, who had been digging into a zabaglione, looked up dreamily. "There was no National Health Service in the twenties, was there?"

"Quite correct, love. I'm glad to hear you're studying social history."

"So how could she afford a sanatorium?"

"I have an idea she told me some relation was stumping up for it," said Mrs. Blayne. "She said she'd send me the address of the sanatorium, but she never did."

"It's a leading question, I'm afraid," said Nigel slowly, "but did it ever occur to you, or anyone who knew her at the time, that she was really going off to have a baby or to get rid of one?"

The J.P. in Julia Blayne came uppermost. "It certainly didn't at the time," she briskly replied. "Not to me, at any rate. And if there was any gossip, I didn't hear it. But of course it's quite possible. Millie was looking very pale. And I remember her being sick the last time I took her out for a picnic lunch on the Common. I put it all down to T.B. then, knowing nothing about it."

"She never mentioned the name 'Protheroe' to you? Or 'Rockingham'? Or 'Geraldine'?"

"Not that I remember."

Nigel had expected nothing else. If Millicent Miles had had an affair with a man at the age of eighteen or nineteen, she would have kept it a clandestine one.

"She used people," Julia Blayne was saying. "That's the basic impression of her that remains in my mind after all these years. She used people, absolutely shamelessly and ruthlessly, as a clever child uses grownups."

"She seems to have used up three husbands," Clare remarked. "And I suppose she used people she knew for characters in her absurd novels."

"Oh God!" groaned Nigel. "Don't tell me I've got to read *them* now!"

"You don't. I have."

"What?"

"Well, the first one. I found a copy yesterday. Published in 1928. All about an innocent young typist who has an illy-jitty by a plausible gentleman of too much leisure and is deserted and ostracized and generally has a rough time till she marries the boss's hard-working son who has always loved her in a solid, silent, respectful sort of way."

"Well, bless my soul! And first novels are usually reckoned to be autobiographical."

Shortly after this, the busy Mrs. Blayne said she must go. Nigel got one or two names from her; but the one he wanted most was useless: Mrs. Blayne, returning to Wimblesham after the war for a school reunion, had heard that the bookshop and its owner had been destroyed by a bomb in 1940.

The whole damned case is a cat's-cradle of dead ends, Nigel said to himself as he made his way to Fleet Street. In the offices of the *Daily Sun*, however, one of his long shots was proved successful. The managing editor, with whom he had fixed an appointment by telephone, had the records searched. The *Daily Sun's* representative in the Wimblesham area, for the big circulation drive of 1924, had been a certain Arthur Geraldine.

"Well, what do you know? He's a top publisher now, isn't he? Got a honey of a libel case coming up—that the man? Never realized he'd worked for us. Before my time, of course."

"Would there be anyone still on the paper who knew him then?"

"Doubt it. We all get fired after a few years, if we don't curl up and die of the horrors first. Wait a minute, though. Old Jackson survived. Was our Advertising Manager. Retired to a life of ill-earned leisure last year. Eunice, get Mr. Jackson's address, will you?"

So much depends, thought Nigel, as he made his way out to Putney to interview the retired Mr. Jackson—so much depends on Millicent Miles' autobiography. How far can it be trusted? How can one discern the objective facts beneath all the subjective, protective coloring she so freely splashed over them? Does that erased "G" stand for Arthur Geraldine? Does she call him "Rockingham" through an association with the Rockingham ware in the cabinet in that terrible little front parlor? If so, was it a chance association or a more deliberate one? Back in the twenties, Stephen Protheroe had rather maliciously remarked, "people didn't always know the value of their own possessions." Suppose Arthur Geraldine, the *Daily Sun* representative, the fanatical collector, had seen the Rockingham set when first he visited the Miles' house, and offered Millicent's father a fiver for it? Mr. Miles was already nearly on the rocks in that summer of 1924; he might well have jumped at the offer, not knowing that the real value of the set was many times greater. Certainly a Rockingham dinner set, of the same delicate apricot color, now graced the Geraldines' table and had been "acquired" some time before their marriage in 1930.

But, though it would have been a sufficiently mean act—a bit of flagrant near-robbery, one could hardly suppose it a strong card for Millicent Miles to play, thirty years later, in a blackmail game; certainly not strong enough to compel Geraldine to kill her rather than be exposed. On the other hand, the purchase of the set might have brought Geraldine into closer contact with the Miles family. He would have been in his early twenties then, and Millicent was a sexually mature girl; if, a little later, he became her lover, one could understand her referring to him as "Rockingham" in her memoirs. Geraldine had certainly become "a power in the world of letters." It was difficult, perhaps, to think of him as ever having been a "shy, gangling young man"; but here, Miss Miles' cliché-ridden style must be reckoned with: for her, all young men might be shy and gangling.

The main snag in this theory was that, according to her autobiography, she first met "Rockingham" a few weeks after her eighteenth birthday. Her birthday was August 3. The episode of the man ("Never trust a man with thin lips") who, she alleged, had made a pass at her, occurred in the summer term, when she was still seventeen. Of course, her memory for dates might well have been defective; or some obscure mental process might have caused her deliberately to postdate the first meeting; or, yet again, if the story of his making a pass at her was a fabrication (wish-fulfillment?), she might only have seen him when he visited her parents' house as the *Daily Sun* representative, and not met him till he came into the bookshop a month later.

Mr. Jackson, a rubicund, white-haired man, received Nigel hospitably.

"Tea or whisky? We always had something in the office at 3:30, and I like to keep up old customs."

Opting for tea, Nigel presently explained, with some caution, the object of his visit.

"Geraldine? Geraldine? Yes, I remember him," said Mr. Jackson, bouncing vivaciously in his chair. "He's come up in the world, hasn't he? When was it now?—middle twenties. Yes. His Lordship, our proprietor"—Mr. Jackson piously crossed himself—"had one of his farcical notions for raising the tone of the paper. Signed on a number of university men. No, not me, I came from the gutter. Geraldine was one of them. Next thing, his Lordship had yet another bright idea—he swarms with ideas, you know, like lice—why not use these highly-educated types to push our circulation campaign? We were offering a complete set of the World's Great Novels, with knobs on, if you took out a year's subscription to the rag, see? Well then, his Lordship conceives this truly Napoleonic notion—send out the university boys to tout for subscriptions. For why?—as we say in our editorials. Because they can unload their book learning onto the prospective customers—tell them exactly why they'll never be really human till they've read the World's Great Novels, complete set bound in elegant linoleum, absolutely free gift in return for one year's subscription. You get the beauty of it?"

Nigel admitted that he did.

"The Oxford and Cambridge chums were properly riled, I can tell you. All they wanted was to write beautiful prose and get by-lines. Instead, the poor sods had to go padding round the outer purlieus giving private lectures on Dustyeffski and who have you to a lot of morons who only wanted to curl up with a nice sizzler by Elinor Glyn. Took the bloom off the Blues in no time, believe you me."

"How did Geraldine get on at the racket?"

"He lived to tell the tale. Only with us about a year, I think. Then he retired gracefully into publishing."

"Did you see much of him?"

"A bit. The noble army of martyrs had to report in at the office twice a week and have their corns attended to. Dour, stocky chap—a slit where his mouth should be—Irish—a bit of a lad with the girls, I'd guess. Though I'd rather be kissed by a shark, myself. Mutatis, of course, mutandis."

"No actual scandal about a girl? One of the customers' daughters?"

"Not that I heard. And if he'd forced his welcome attentions on a young female, it'd have been all over the office before she'd had time to renew her lipstick."

"Would it have been part of his job as a representative to visit the local bookshops?"

"Bless your heart, no. We were in competition, don't you see? Course he might have gone in to buy a book—never know what these university types aren't capable of when roused."

"You say they all had literary ambitions. Do you remember if Arthur Geraldine wrote at all then?"

"Now you're asking something. Couldn't be sure, but I've an idea he had a piece or two in the hoilier-than-thou weeklies. But his real hobby was collecting bits of china. Probably that's why he stuck the job as long as he did. While he's explaining the benefits of the *Daily Sun* free gift scheme, he can run his eye over the mantelpiece and see if the client's harboring a Dresden angel unawares. Wimblesham was more countrified those days, and the natives wouldn't always know the value of the sweet little vases Great Aunt Flossie had left them. He picked up quite a few bargains, I know for a fact, in the crockery line. Remember him panting back to the office one day with a dinner plate in his bag—one of a set he'd chiseled some sucker out of—not that he put it that way; you should have heard him crooning over it! Looked like any other dinner plate to me."

"What color?"

"Oh, a sort of washed-out brown. Browny-yellow."

"Apricot?"

"Fair enough."

Nigel's mood of elation, on leaving the ebullient Mr. Jackson, was short-lived. It seemed certain that Arthur Geraldine had "acquired" the Rockingham set during his house-to-house canvassing of Wimblesham in 1924. It seemed likely, but was in no way proved, that this set had belonged to the Mileses. If, as a girl, Millicent Miles had known about the transaction; if, last summer, she recognized Arthur Geraldine as the quondam *Daily Sun* representative who had bought the set at far below its true value, this would explain both Geraldine's unwillingness for Nigel to start digging up the remote past, and the unusual concessions he had made to Miss Miles. She might well have been putting pressure on him. But the affair of the Rockingham set was still nothing like adequate motive for murder.

And what possible link could there be between this and the dirty work with the proof copy of *Time to Fight?*

If, on the other hand, Millicent Miles had had a child by Rockingham-Geraldine in 1926, if he had abandoned her and the child, then she would have a far stronger blackmailing card against him, and he a proportionately stronger motive for murder. But why had she not played it years ago? She had moved in the literary world since about 1930; she would be bound to know that Arthur Geraldine had become a successful publisher. Perhaps the answer was that only recently, when her stock as a writer had slumped, would she need to apply pressure on Geraldine.

The child, Nigel reflected as the underground train rushed him back to Kensington, would be thirty now. What had happened to him, or her? Not that there was the least evidence he had ever existed—only a tenuous hint given by the page the murderer had inserted in the dead woman's autobiography—the(deliberately misleading?)statement that she never had a child by her first lover.

Nigel's eye was caught by a headline in the newspaper a man opposite him was reading. NATIONAL SERVICE TO BE CUT? Some of the young soldiers destroyed in the "holocaust" of the Ulombo barracks, he recollected, were National Servicemen. A spark leaped a gap, and the two dark poles of the case were suddenly linked in Nigel's mind. The Ulombo barracks affair had taken place in 1947, when Millicent Miles' child would have been twenty-one—National Service age.

And was there ever a more hare-brained conjecture than that? thought Nigel as he came out into Kensington High Street—a cobweb theory built upon a mere coincidence. Of course it must be a coincidence.

Nevertheless, he was walking very fast indeed up Campden Hill Road, impatient to get to the telephone.

When he had let himself into the flat, his housekeeper told him that a young gentleman had arrived an hour ago, asking to see him. A Mr. Gleed. She had told him that Mr. Strangeways might not be back for some time, but he said he would wait. Mrs. Anson was clearly a bit flustered: strangers who arrived unannounced and refused to go presented a problem she could not cope with. Assuring her that no harm had been done, Nigel entered the sitting room.

Cyprian Gleed had made himself at home. He was sitting in Nigel's armchair, with a glass of Nigel's whisky at his elbow, and his mother's autobiography on his lap.

"Good evening," said Nigel coldly. "I see you found the decanter. Wouldn't you like some water with it?"

"No thanks." Cyprian Gleed seemed unsnubbable. "I didn't mean to stay so long, but I couldn't resist this horror comic," he added, indicating the typescript on his lap.

"I see. Well, excuse me a minute." Nigel went into his bedroom and picked up the extension telephone there. General Thoresby, he was told, would be home shortly. Nigel left a message, asking the General to ring him as soon as possible.

The purpose of Cyprian Gleed's visit was not immediately apparent. He complained of being incessantly questioned and followed by policemen. He resented being told nothing about the progress of the investigation. Finally, by a devious route, he came to the point. When could he expect to inherit his mother's estate?

"Why ask me? You should deal with her solicitors."

"Oh, they just say there are the usual formalities to go through. I don't understand all their legal jargon. What formalities?"

"Well, they have to make sure there's no will, and that you are the next of kin."

"But there's no question of that." Gleed's red lips twisted behind the black beard, as he jabbed a finger at the typescript. "She seems to have had some squalid love affairs at a tender age, but she makes it quite clear the union was unblessed by fruit. And she certainly didn't have any other children by her various legitimate husbands."

"You'd be fairly safe to borrow on your expectations, then."

"You think so? Do you lend money, by any chance?"

"Not to you."

Gleed took another drag of his neat whisky. "I suppose Wenham & Geraldine aren't going to publish this tripe now?"

"Why not? It's unfinished, but—"

"Good God! Have you read it? The meanderings of an imperfectly sublimated nymphomaniac! How the hell should I ever live it down?"

"What do you care? It'll bring you in a lot more money. That's all you're interested in, isn't it?"

"Ah, you're shocked. Money's a dirty word—never mentioned in your high-minded circles," jeered Cyprian. "Whereas mothers are holy, however bitchy—"

"You really are a contemptible little creature," Nigel broke in, speaking with icy deliberation. "You talk like a permanent adolescent, and behave like a spoilt child. Do you really think you impress anyone by mouthing all this vicious muck about your mother? The trouble with you is that you've got no talents and no charm—you're a total failure as a human being, and you know it—so you've got to compensate by making a pretentious nuisance of yourself all round. Well, did you cut your mother's throat or didn't you? There are no witnesses present. You've a chance to make yourself really interesting for once, and with impunity. Come on, my bold little Orestes."

Nigel's outburst, though sharpened by the sheer physical revulsion he felt for Cyprian Gleed, was a calculated experiment. How would Gleed respond to treatment even more outrageous than his own utterances?

Cyprian Gleed stared back venomously at Nigel. "Have you quite finished your orgasm?"

"You've often wished your mother dead. But you'd neither the brains nor the nerve to kill her. That's why you spent so much time shooting off your poisonous tongue at her, from a safe distance. Well, she's dead now. So you can stop shooting."

"Ah, I thought the *de mortuis* stuff was due. That is really more than I can face. I must be going. Sorry to leave you in uncertainty about my matricidal history; but when next you try a bit of psychological vivisection, I'd recommend you to use a scalpel, or even a razor, not a blunt flint—it was all too painfully obvious." A singular, gloating expression came and went on Cyprian Gleed's face. "I find you quite remarkably antipathetic. Do you mind if I use your lavatory to be sick in?"

"First door on the right down the passage."

As Cyprian was going out, the telephone bell rang.

"I'll let myself out, when I've finished," Cyprian said. "You probably won't be seeing me again."

The door closed behind him, and Nigel took up the receiver.

"Strangeways? Thoresby here. You asked me to ring you."

"Yes. That Ulombo barracks affair. Have you got the casualty list?"

"Just a minute. I'll dig it out of my files. They were all scuppered, y'know. Hold on."

When the General was back on the line, Nigel asked, "Can you tell me, is there a Geraldine on the list, or a Miles?"

"Now what the devil are you up to, my dear fellow? I don't remember a— I'll look."

The twenty seconds of silence seemed twenty minutes to Nigel. At last General Thoresby said, "No. Drawn a blank there, I'm afraid."

Nigel felt the disappointment like a kick in the belly. Yet his had been a pretty wild shot in the dark.

"Very good of you, sir. I'm sorry to have bothered you. Just another theory of mine miles off the target."

"Wait a bit, my boy. If you're interested in literary names, publishing names and so forth, what price Protheroe?"

"Protheroe?"

"Yes. Isn't that the name of Wenham & Geraldine's reader?"

"Yes, but—"

"Well, there's a Paul Protheroe on this casualty list. A corporal. One of the National Servicemen who was on the strength there."

XII *LEAD IN*

While Nigel Strangeways was delving into the remoter past, Inspector Wright and his team had been investigating the more recent. At a conference the morning after General Thoresby's telephone call, the two men pooled their information. Double-checking their movements of the previous Friday, Wright had finally eliminated from suspicion such peripheral figures as the General himself and the dead woman's third husband: the latter had had no personal contact with her since their divorce. As to the Wenham & Geraldine people, and Cyprian Gleed, Wright's inquiries had still produced mainly negative results.

It was clear to Nigel that the Inspector had been concentrating upon Cyprian Gleed. Naturally so. There was no love lost between him and his mother; he was an irresponsible, amoral and possibly violent character; he needed money badly, but his mother had refused to give him any more until his next allowance was due. Wright had discovered, furthermore, that Gleed was being pressed hard by several creditors for payment of quite considerable sums. So far, so good. But house-to-house inquiries, both in Angel Street and the vicinity of Gleed's flat, together with the usual appeals to taxi drivers, etc., had so far turned up no trace of evidence that Gleed had left his flat between 4:30 and 7 P.M. on the night of the murder.

Repeated questioning had done nothing to shake the alibis, such as they were, of the other suspects: they had not, any of them, materially modified the evidence given in their original statements, nor contradicted themselves over matters of detail.

It seemed almost certain to Wright that the murderer must have brought a bag with him in which to take away the weapon and his bloodstained clothing. The question was, when did he remove the bag from the building? It could have been done either (a) immediately after the murder, or (b) at some time during the weekend by anyone who had a key to the side door; it would have been quite safe to leave the bag overnight, since the office part of the building was empty from Friday evening till Monday morning. Inspector Wright's research into this postulated piece of luggage had yielded the following results:

Stephen Protheroe had brought a kit bag into the office on Friday morning. Miss Sanders noticed he was carrying it when he left the office at 5:20 P.M. The Hampshire friends, with whom he stayed the weekend, said it was his first visit and confirmed that he had brought a kit bag, of the same color and size. Moreover, his host had been chatting with him in the bedroom when he unpacked it: Stephen had thrown the contents onto the bed—an unthinkable thing to do if they included bloodstained clothing and weapon.

Basil Ryle. Nobody had seen him with a bag that Friday. On the other hand, no one could swear he had not brought one to the office, and it might easily have been hidden in his room. He had walked across Hungerford Bridge to the Festival Hall; he could have checked a bag in at a cloakroom there, before dinner, but there was no evidence that he had done so.

Liz Wenham was definitely not carrying a bag when she came to the cocktail party in Chelsea. Her taxi driver had been traced, and stated that she had no luggage with her when she entered his cab in the Strand shortly after 5:30. They had not stopped anywhere on the way to Chelsea.

There remained Arthur Geraldine. His wife said that she had heard him come into the flat at about ten to six. He had gone straight to the bedroom, bathed and changed—his normal custom. He could have taken off bloodstained clothes then and concealed them, but it would have been a risky proceeding. He was his usual self at dinner, so Mrs. Geraldine said.

As to the weekend, the situation was even more open. Arthur Geraldine, for instance, had worked at home on the Saturday morning, then gone for a spin with his wife to Richmond. The garage attendant stated that Mr. Geraldine was carrying no bag when he arrived to fetch the car. The whereabouts of Miss Wenham and Mr. Ryle could not be corroborated for the whole of Saturday and Sunday. Stephen Protheroe had unquestionably been in Hampshire. Cyprian Gleed's drunken progress would be almost impossible to trace from beginning to end.

But Inspector Wright was now skeptical of the theory that the murderer had gone back to retrieve a bag during the weekend. Angel Street would be much quieter then, and he would run a greater risk of being noticed letting himself in at the side door. Also, he could not be sure that Mr. and Mrs. Geraldine, or their maid, might not be going out when he entered.

"So we're back in the bosom of dear old Mother Thames," said Wright, raising his eyebrows interrogatively at Nigel.

"I see. Yes. It looks the best bet. Not the Embankment, though."

"Not unless he lost his head and tried to get rid of it as soon as possible. No. Hungerford Bridge. I took a stroll across it last night in the rush hour. Plenty of people hurrying over to catch trains at Waterloo. Like a crowd of sheep, these commuters—never notice anything—just want to get back to the little woman and the telly. As you walk over the bridge south, there are small arc lights on the girders above you, to your right. Your left side is in shadow. You hold the bag over the rail with your left hand and just let go, without stopping walking; or you stop a moment to admire the beautiful reflections of light on the water. It's just too easy. There are trains rumbling over the bridge almost every minute during the rush hour. Drown the splash. Good-by to the evidence. Of course, you'd have to weight the bag to make sure it sank. So it'd have to be a pretty strong one, or it'd burst on hitting the water. But no doubt he'd give his mind to a little problem like that."

"You'll drag the river? At low tide? Or use frogmen?"

"It's being done. But the tides are fierce. Bag's halfway to Tilbury by now, I should think. Scuttling along the river bed."

"Well, if you're right, it narrows down the suspects a bit."

"Yes. Protheroe and Ryle both walked over the bridge between 5:25 and 6:15."

"And told us, without any hesitation, that they'd done so."

"Exactly. Whereas your engaging boy friend, Cyprian Gleed, says he was at home waiting for his mum to turn up, but could have been prancing about on the bridge."

"Are you issuing an appeal to the public?"

"Yes. It'll be in the evening broadcasts, and in tomorrow's papers. And the wires'll be red-hot with people ringing up to say they thought they might have seen a man or heard a splash or something. But as for identification—I ask you! Turn facing the river, lean over the rail, and the lights are behind you; your face is in shadow. Height, sex, and possibly clothing—that's all anyone could see. But these commuters wouldn't notice if you threw an elephant over."

Nigel now gave Inspector Wright a summary of his own operations over the last thirty-six hours. The trap he had laid at dinner with Arthur Geraldine and the others had remained unsprung: no attempt had been made to steal the typescript of Millicent Miles' autobiography—not that it would have benefited the murderer, in fact, to do so, for there were photostat copies of the relevant pages at Scotland Yard. Cyprian Gleed's interest in the book, however, gave Wright food for thought, and he made Nigel repeat the whole of their conversation.

"So he wanted to know about his mother's estate—the legal aspects," commented Wright thoughtfully. "And he said there was no question of him not being the next of kin, because, though she'd had a love affair as a girl, there'd been no child by it. That's rather significant, isn't it?"

"Why?"

"It suggests that Gleed doesn't realize that, even if she had had a child then, it couldn't be next-of-kin—bastards don't inherit where there is a legitimate child."

"All this is assuming she had a child *and* it was illegitimate." Nigel went on to tell Wright of his discoveries about Arthur Geraldine—Wright did not appear much interested in these—and the existence of a Paul Protheroe who had died in the Ulombo barracks.

"General Thoresby is having his army record looked up for me, and I've got a chap at Somerset House searching the records of births for 1926. He's going to ring me here if he finds anything before eleven o'clock."

"Protheroe's not all that uncommon a name."

"No. But if she had a child, and Stephen Protheroe was its father, we at last get a link between the murder and the tampering with Thoresby's book."

"You mean, either Protheroe, or Miss Miles, or both of them together, fiddled with the proof copy, knowing the boy had been killed at Ulombo—?"

"Exactly. But it was Protheroe, I'd say, who did the fiddling. He'd want Blair-Chatterley exposed, to avenge the death of his son, caused by Blair-Chatterley's incompetence. Miss Miles discovers he has tampered with the proof, threatens to tell on him to the partners; he kills her to safeguard his job."

Nigel enlarged upon his theory with considerable excitement. When he had finished, Wright sat back scrutinizing him, head cocked like a bird over a wormcast.

"Well, well. It's all very pretty. On paper. And you believe Protheroe is capable of murder? *That* sort of murder?"

Nigel was spared answering this uncomfortable question by the ringing of the Inspector's telephone.

"For you. Somerset House."

Nigel took the receiver, listened intently for a minute.

"Hold on," he said; then, to Wright, "A child was born on November 29, 1926, at Greengarth in Northumberland. Paul Protheroe. Names of parents on birth certificate—Stephen Protheroe and Millicent Protheroe. And it *is* the Paul Protheroe who was killed at Ulombo: date of death fits. Any questions?"

"Ask him to search for record of a marriage between Protheroe and Millicent Miles. And ask him if anyone else has been making similar inquiries—within the last six months, say."

Nigel passed on these requests, thanked his informant and replaced the receiver. Inspector Wright, who could never keep still for long, was doing a tap dance with his fingers on the desk.

"Well, it's your pigeon," he said. "You're investigating that libel affair for the publishers, and now you've got a lead. Chase Stephen Protheroe over that as hard as you like. But lay off the murder angle, Mr. Strangeways. Frankly, I'm not with you there—we haven't a shred of evidence yet."

Ten minutes later, when Nigel arrived at Angel Street, Miriam Sanders told him that Miss Wenham wished to see him immediately. Deferring the showdown with Stephen Protheroe, he went to her room. It was sunny this morning. The light, striking in obliquely through the fine, tall window, enhanced a brightly colored design for a dust jacket lying on the work table and picked out the gold lettering on the spines of a uniform edition in the top shelf of a bookcase. It was less flattering to Liz Wenham's looks. Her gray hair, even her eyes, usually so animated, lacked luster today; the rosy-russet cheeks were sallower; she had the etiolated appearance of a fresh-air woman confined too long indoors.

"Ah, here you are," she greeted him, making it—as often she did—sound like "where have you been all this time?"

"You're looking a bit tired," said Nigel sympathetically.

"So would you be if you had to run this firm practically

single-handed. Arthur's tied up with the libel business half the time. And Basil seems to have conked out completely— can't keep his mind on his work at all. Damn that woman! What I'd do without Stephen I just don't know."

"He's bearing up all right?"

"Stephen doesn't cave in. He's got backbone."

"Never?"

"I don't think he's missed a day, holidays apart, for ten years." She suddenly brandished a sheaf of papers in Nigel's face. "How many times have I told you to type the subscription orders in a separate—?"

"Yes, Miss Wenham. I'm sorry."

The secretary, who had entered silently, took the papers and fled from the room as if blown out of a wind tunnel.

"Silly girls! They're all flustered out of their wits. When is that Inspector of yours going to take himself off our necks?"

"When he's made an arrest, I suppose."

For an instant, there was a glaze of fear, or anxiety, over Liz Wenham's clear eyes.

"It's all nonsense," she roundly declared. "One of us! Quite preposterous. Is the Inspector a shoe fetishist?"

"I beg your pardon?"

"The other evening he came round to my house and started poking about in my shoe cupboard," said Liz with high indignation. "It's enough to turn me morbid."

"He was looking for galoshes, I expect."

"*Galoshes?* Gracious heavens, I've never worn such things in my life! Why galoshes?"

"The murderer wore a pair. To prevent his shoes getting wet," Nigel flatly replied.

"Oh yes, I remember you mentioned it at dinner. But does your Inspector really suppose that the—that whoever did it would just put the galoshes back in a cupboard? Piffle. What was I saying before that idiotic girl—?"

"You were saying that Stephen Protheroe hadn't missed a

day for ten years. He doesn't look all that strong to me. Is he never ill?"

"He did have some sort of a breakdown about ten years ago—" Liz Wenham uttered the word "breakdown" in a markedly derogatory tone. "Away for a week or two. But there's been no bother of that sort since."

"Overwork, was it?"

"Nobody breaks down from overwork. Some emotional trouble, I fancy." Liz Wenham's tone made it clear that she had little more sympathy for emotional troubles than for the breakdowns they caused.

"Would that have been in 1947?"

"I can't think what possible interest this can have for you. Still. 1947? It was the year we published *The Buried Day*— that I do remember. Stephen was doing the editorial work on it. Held up production, you know, his being away." Liz leaned back and took a volume from the shelves behind her. "Here we are. Yes, it was 1947."

Nigel now reminded Liz Wenham that she had asked to see him.

"Oh yes. Two things. I wish you'd have a word with Basil; he's gone broody as I told you; I can't manage to get him out of it. Needs a tonic, I daresay. And what was all that talk the other night," she continued, glancing shrewdly at Nigel, "about Arthur's Rockingham set? You seem to have got under his skin, somehow."

"Do you mean the subject cropped up again after I left?"

"No. But when you said something about a Rockingham mentioned in the Miles autobiography, I noticed that Arthur looked annoyed. No, after you'd gone—I should tell you quite frankly—Arthur raised the question of paying you off. We're very grateful to you for holding our hands through this police inquiry, but—"

"I do quite understand. There's the *Time to Fight* trouble to be cleared up, though."

"I thought you'd come to a dead end over that."

"I had. But yesterday I gathered some new information. I believe I know now why the proof was tampered with. And the 'why' points to the 'who.' "

At this point, Liz Wenham's telephone rang and she went into a clinch with the literary agent at the other end. It would evidently be a prolonged tussle, so Nigel got up and made his way to Basil Ryle's room.

The young man was in bad shape, no doubt of that—staring into vacancy and seeming, for a moment, not to recognize Nigel. His eyeballs, turning painfully to his visitor's face, moved as if they had lead weights on them.

"Sorry you're under the weather," said Nigel. "Couldn't you take a few days off?"

"How much longer is this going on, for God's sake?" said Ryle in a collapsed voice.

"The investigation? I've no idea."

"I didn't mean that."

"As long as you refuse to face up to reality," Nigel coolly said.

With his red hair tousled and dank, his defenseless face, Basil Ryle looked like a sick schoolboy.

"I've faced too much reality this last week; I never want to see it again."

"No you haven't. You're still trying to make excuses for the way she treated you, trying to pretend to yourself that she was the marvelous creature you once thought her. . . . No, listen to me. Do you know what you're really fretting about? Not the loss of her. The loss of your self-respect. Self-respect means a lot to a chap who's come up the hard way, as you have. And yours was too much involved in your relationship with her. When she turned on you and showed herself up in her true colors, she dealt a mortal blow at your ego. She falsified all the familiar landmarks of your relationship, so of course you were lost. If she is not she, who on earth am I? It's a disintegrating experience, to feel like that. And ever since, you've been trying to recreate your illusion of her, because it's the only way you know to restore your self-esteem and recreate your self."

"I expect you're right," Basil pathetically murmured.

"But it's all wrong! You can't rebuild your life on an illusion. It gives way beneath you all the time. That's why you're in this wretched state now—it'll turn neurotic if you aren't careful. You've got to accept it that you were a fool, and she was—well, what she was. Stare *that* in the face for a while, and you'll be cured."

Basil Ryle lifted his lugubrious, red-rimmed eyes. "My God, I was a bloody fool all right! I ought to be—"

"O.K. O.K. Face it, but don't wallow in it. Self-disgust is a good stimulant but a bad habit."

"Sententious bastard, aren't you?" said Ryle, but smiling for the first time. "Funny, I didn't much take to you when we met—ten days ago, was it?—time's been no object since—" His voice trailed away, and the look of being saturated in misery returned to his face.

"Do you believe in brainstorms?" he asked.

"People have them."

Making a great effort, as if dragging out the words against their own will, Basil Ryle said, "That's what terrifies me—that I may have done it myself. . . . I could have *killed* her, the night before; I ran out of her house so that I wouldn't lay hands on her. I was in an absolute muck sweat. How do I know I didn't have a brainstorm the next night? All I remember is sitting here, in a daze of misery, humiliation, fury, and then going off to try and soothe my savage breast at the Festival Hall."

"Now do pull yourself together. I doubt if you're a schizophrenic. Even if you are, and could have killed her in what you call a 'brainstorm,' you certainly wouldn't have planned everything ahead, as it was planned. This was a premeditated crime, thought out in every detail."

"Thank God for that!—Well, you see what I mean. You've certainly taken a load off my mind."

"You couldn't have killed Millicent Miles in a brainstorm," said Nigel, gazing straightly at Basil Ryle.

"Yes, I realize that now. . . . Oh, I see. I'm not a madman but I may be a fiend?" Ryle laughed unconvincingly. "You're a disturbing chap, aren't you? Only the other night you were saying the scene I had with Millicent pointed to my innocence."

"It could point two ways, like most things. A subtle murderer might go to her house and deliberately provoke the scene—we've only your word for what happened, and you had the ideal supporting cast—a nosy German girl who could overhear a quarrel, but couldn't understand English. So the policemen reason, if he was going to kill her, he'd have killed her then, under intolerable provocation; but he didn't; therefore he's not likely to have killed her the next day."

"That's all too subtle for me," said Ryle; then, with sudden, half-humorous pugnacity, "Now for God's sake run away and talk to somebody else. I've got work to do."

Thoughtfully, Nigel made his way along the passage toward the stairs. There was something at the back of his mind, just out of reach; it would have to wait, whatever it was; for there remained the more pressing problem of Arthur Geraldine to be cleared up, and Nigel did not know how best he could deal with it. He realized, too, that he was postponing the showdown with Stephen Protheroe. The worst thing about his job, which he found continuously absorbing, was that he got to like so many of the people it brought him in contact with. And some of them were murderers. And murderers, by and large, tended to be less abominable than their crimes.

At the foot of the stairs he turned round again and marched resolutely back toward Arthur Geraldine's room. Carefully modulated aggressiveness should be the note. Now look here, Geraldine, do you or don't you want this proof-copy problem solved? Miss Wenham tells me you don't wish to avail yourself of my services any longer. Very well. But the problem of the proof copy is now all but solved, and I presume you would still care to know the identity of the miscreant. What has hampered my investigations throughout is the refusal of certain persons to admit that they knew Miss Miles in the past—two persons, Geraldine, of whom you are one.

Rehearsing this agreeably theatrical speech brought Nigel to the senior partner's door. As he opened it, he was almost driven back by a cloud of smoke and a wave of stuffy warmth. The room, he soon deduced, was not on fire—or, if it was, the persons sitting at the long table were taking it remarkably calmly: six prepossessing young gentlemen, equally distinguished in dress and in feature, all smoking away like fury. Nigel's catarrh and sinus trouble had deprived him of his sense of smell. At the head of the table, Arthur Geraldine looked round.

"Oh, good morning, Strangeways. Anything urgent? I've got this travelers' meeting till lunch. I don't think you've met these gentlemen." Courteous as ever, Geraldine performed introductions, the six distinguished-looking persons rising to their feet as one man, and greeting Nigel in turn with a "Good morning, sir"—a ceremony somewhat marred by Liz Wenham, who breezed in at that moment, exclaimed "My God, what an appalling fug!" and threw open the nearest window.

From the foot of the long table, Stephen Protheroe winked at Nigel.

"These gentlemen," pronounced the senior partner with considerable grandeur, "are the spearhead of the firm's fortunes—our panzer brigade. They not only travel hopefully, they also arrive."

A gratified murmur from the panzer brigade greeted this trope, and they sat down again as one man. Overcome by a mounting sense of unreality, Nigel made his excuses and left the conference. Arthur Geraldine's public manner was so different from his easy, unbuttoned private one. Climbing the stairs, Nigel reflected upon the paradoxical Anglo-Irish alternations of boisterousness and ceremoniousness, hauteur and horseplay, glibness and reticence.

Alone in Stephen Protheroe's room, he flipped through a few papers, stared out of the window, glanced in at the room where Millicent Miles had met her death, empty now except for the table and one chair. It was odd to think of her and Stephen working next door to each other all those weeks, with nothing between them but a sliding window and a dead bastard. Or had Paul Protheroe been legitimized? Nigel rang Somerset House. After a few minutes' delay his friend came to the telephone: no, there was no record of a marriage between Stephen Protheroe and Millicent Miles, either before or after Paul Protheroe's birth. Whoever had registered the birth must have given a false name for the child's mother, in the interests of respectability; unless, of course, there had been two Millicents in Stephen's past—a possibility too discouraging to contemplate. As far as Nigel's informant knew, there had been no other inquiries recently about this birth certificate.

At a loose end now, Nigel began to poke about in the drawers of Stephen Protheroe's desk. One of them was locked. Unable to find a key in any of the other drawers, Nigel set to work and picked the lock. Presumably the police had searched this desk some days ago; yet Nigel felt a certain tension as he worked. The drawer did not contain a razor, galoshes, bloodstained clothing. But it was not empty: Nigel saw a litter of paper—loose sheets, some of them yellowish with age, all of them dusty. He took them out. Beneath them he found a book: it was *Fire and Ash*, the poet Protheroe's only living child.

There had been many other children, though—born dead or strangled at birth, weaklings, monstrosities. The loose papers under which *Fire and Ash* was buried proved to be work sheets of poems. It did not take Nigel long to discover that they were unpublished poems, not earlier versions of those which had appeared in *Fire and Ash*. They were undated; but it seems probable they had been composed after the publication of that book—why should Stephen preserve juvenilia written before 1927?

As he studied the work sheets, Nigel had a recurrence of the feeling experienced in Cyprian Gleed's flat—a feeling compounded of pity, embarrassment and shuddering distaste; the drawer from which he had taken these sheets was a mausoleum of dead ends. There were dozens and dozens of poems, hatched and cross-hatched with alternatives in Stephen's neat, spidery writing, and not a single one of them was finished. The emotional conflagration which had produced his masterpiece must have suffocated Stephen's genius beneath the weight of its own ashes. These manuscript poems were lifeless. For all his technical skill, Stephen had been unable to infuse them with even the similitude of life. Nigel imagined him beginning each new one with ever-diminishing confidence, working his way painfully toward some half-glimpsed objective, then losing sight of it, losing touch, losing interest.

Replacing the pitiable fragments in the drawer, Nigel turned with relief to *Fire and Ash*. It was not the first time he had reread it during the last ten days, but it had lost none of its impact through familiarity: lines rose up again and jolted him under the heart with sickening violence and precision. A raw, harsh work, tracing a man's love for a woman through all the stages from passionate faith to savage disillusionment—critics had compared it, when first it was published, to *Modern Love* —this sequence moved, without pity or self-pity, to its tragic close. Great images of love and lust, of innocence, betrayal, hatred and despair flared up from the pages, shedding a light ever colder and more relentless upon the human situation, so that the man and woman in the poem were dwarfed by the lengthening shadows of their destiny.

Nigel closed the book, put it back, and locked the drawer. A quarter of an hour later, when Stephen Protheroe returned from the conference, Nigel stared at him for a few moments, as if he could not recognize in this shrimp of a man the creator of that merciless, passionate masterpiece, *Fire and Ash*.

"Anything wrong?"

"We've got to have a talk," said Nigel. "About Paul Protheroe."

XIII *TRANSPOSE*

Stephen Protheroe's reaction was not what Nigel might have expected. After regarding Nigel meditatively for a few seconds, he gave a little, sad, abstracted smile, and said, "Ah well, I suppose it had to come out."

"Paul was your son, by Millicent Miles."

"No, no, you've got it all wrong."

"But it's registered at Somerset House—"

"I know. . . . Look here, let's go and have something to eat. I can't confess on an empty stomach." There was a glint of mischief in Stephen's eyes. He refused Nigel's invitation to a restaurant, saying that such places gave him claustrophobia. "Come back to my flat and I'll run you up an omelet." Then, as Nigel hesitated, "I'm a tolerable plain cook. And if you don't care to eat a man's salt when you're accusing him of a dark deed," added Stephen with a grin, "you can damn well eat your omelet without salt."

There was no refusing this. Stephen Protheroe appeared to have taken control. In the taxi, as though Nigel must be relieved of any embarrassment, Stephen gave a droll account of the travelers' meeting; he showed no sign of awkwardness, let alone guilt; indeed, he was more vivacious than Nigel had ever known him—from relief, perhaps, that he would at last be unburdening himself of his secret.

Nevertheless, when they were in the neat little Holborn flat, Nigel did not let his host out of his sight. He had no intention of being caught napping. There were other weapons than razors, and poison was one of them, absurd as it seemed now to connect the amiable Stephen with such things. Nigel followed him into the kitchen, saying,

"Do you mind? I'm everlastingly curious about the way other people live."

Stephen Protheroe lived, it was at once plain, with remarkable efficiency. His kitchen was spotless, well equipped with labor-saving units, everything in punctilious order and ready to hand. Stephen himself set about preparing the meal with a concentration that won Nigel's approval. Breaking the eggs, he said over his shoulder,

"You'll find some wine in the cupboard by the door. Let's have a bottle of hock. Run the cold tap over it for a bit, will you—not what the pundits would approve, but I haven't an ice bucket. Corkscrew in the left-hand drawer of the dresser."

Nigel followed the instructions, still keeping his eye upon Stephen, who seemed much too absorbed in his work to notice that he was being watched.

"I must say you've got a high-class *cuisine*," Nigel remarked.

"Well, I like looking after myself and I enjoy cooking. After Paul was killed, I was able to launch out and indulge myself a bit." The surprising little man turned round, handing Nigel his omelet on a plate. "This won't poison you, at any rate. Go and eat it while it's hot. Through there. You'll find the table laid. I'll bring in the wine and some bread and butter."

The sitting room too was scrupulously tidy. There was a place laid for one at the round table, and a vase of freesias on the mantelpiece; comfortable armchairs, long low bookshelves painted white, a reading stand, a crimson carpet, a Matthew Smith blazing with color over the fireplace.

Nigel strolled across and examined two photographs on the mantelpiece: a small boy riding a tricycle, and a young man in military uniform.

"Yes, that's Paul," said Stephen, who had entered silently. "Do come and eat your omelet. Unless you prefer it leathery."

Meekly, Nigel sat down and began, while Stephen set a place for himself and poured out the wine.

"Your health," said Stephen. "How's the sinus trouble?"

"Not much better, I'm afraid."

"Salt?" Stephen's voice, at its most resonant, made a challenge of the word; his fine eyes held Nigel's.

"Thank you, I will. Without prejudice."

Not till they had eaten the second course—fresh fruit from a bowl of Italian pottery—was the subject in both their minds mentioned. Nigel had his own reasons for wanting Stephen to introduce it again; but Stephen seemed in no hurry to do so, preferring to talk about his flat, of which he was pleasantly proud—he was so comfortable here, he said, and enjoyed his own company so well, that he very seldom went away. "Solitude suits me. Solitude in a crowd," he added, with a gesture toward the enveloping city. "Are you married?"

"I was. My wife is dead."

"I like women. But I don't want them around all the time; they demand too much attention, and what's worse, they pay one too much—you know, watching every expression that passes over one's face—is he getting tired of me? Or is it just a twinge of his indigestion?"

Nigel laughed. "You never thought of marrying Millicent Miles then?"

"Great Scott, no! Even at the time, she made me sick."

"At the time when she had your child?"

Stephen shot him a swift look, then motioned him to one of the armchairs. "Coffee for you?"

"No thanks."

"Well, let's get started." Stephen rubbed his hands briskly. "Tell me how you discovered about Paul."

"I got the impression early on that there was some long-standing relationship between you and Miss Miles. Then her autobiography spoke of a love affair she'd had, at nineteen, with a man who subsquently deserted her."

Stephen drew in a sharp breath, as if he had been stabbed,

and his face contorted. "*Deserted her!*" he exclaimed. "Never mind. Go on."

"She apparently told Ryle that she'd been seduced when a girl, and had a still-born baby."

Stephen looked puzzled. "But is there a reference in the autobiography to her having a child?"

"Not exactly. Reading between the lines, though—" Nigel broke off. This passage had to be handled with the utmost delicacy. If Stephen was the murderer, it was he who had substituted at the end of the chapter the page which stated that Millicent had not had a child by her early lover; and if this were so, Nigel's disregarding the information given by this page would show Stephen that Nigel believed it to be a fake, and put him on his guard. On the other hand—still assuming that Stephen had committed the murder—he could not, without betraying himself, reveal his knowledge of the substituted page.

"I was trying to find a motive for the stetting of the libelous passages in *Time to Fight*," Nigel resumed. "It had clearly been done in order to damage either the publishers or the author or General Blair-Chatterley. I couldn't find any adequate reason for anyone wanting to damage General Thoresby or your firm, and I was totally bogged down. Then, after I read the autobiography, it occurred to me that, if Miss Miles *did* have a child in 1926, it would be twenty-one in 1947—the year of the massacre at the Ulombo barracks. I rang General Thoresby and asked him to look through the casualty list. One of the names on it was Paul Protheroe. Somerset House did the rest."

"I see. Yes." Stephen seemed to lapse into abstraction.

"And today Miss Wenham told me that you'd had some sort of a breakdown in 1947," said Nigel gently. "The shock of your son's death—"

"*Oh, but Paul wasn't my son.*" Stephen's face was averted.

Nigel stared at him. "What? But the records at Somerset House—"

"Damn Somerset House!" Stephen got to his feet, gazed for a moment at the photograph of the child on a tricycle, then went over and sat by the window. "It wasn't my secret," he slowly began. "That's why I've been keeping my mouth shut. And then, of course, I didn't know what sort of trouble I might get into if it was found out that I'd made a false registration."

"Of Paul's birth?"

"Yes. But I'd better tell you the story from the beginning. I did meet Millicent years ago, when she was seventeen or eighteen. Quite by chance. In a bookshop at Wimblesham. Is that referred to in her book?"

"Yes. But she doesn't give your name—just mentions a man who encouraged her in her literary efforts. She calls him 'Rockingham'—a verbal association with another man she'd met about the same time."

Stephen Protheroe leaned back on the window seat, relaxing a little. "She had talent then, of a sort. I was too young to realize that her heart was made of—of synthetic rubber. Well, I had a brother. Peter was two years younger than myself, and I'd always been protective and auntyish toward him. In 1925 he was at a theological college in Oxford, training for the ministry, and I was struggling along with free-lance literary journalism. During the vacations, Peter used to share my rooms in London. Millicent turned up one day to ask my advice about a story, and she met him there. That's how it all started."

"You mean, Paul was your brother's child?"

"Exactly. Peter was a fine young chap. He had a very strong sense of vocation—and the passionate nature that so often goes with it. He was also, as far as women went, totally inexperienced, repressed, absurdly romantic—the ideal prey for a girl like Millicent. No doubt she had a tough struggle with his religious convictions—she'd have enjoyed that—but she succeeded in seducing him."

The sad anger on Stephen's face deepened, and he pounded the window-seat cushion with his fist.

"You'll see before long why I loathed that woman. Peter came back for the Easter vacation the next year in a terrible state. He hadn't dared to write to me about it, but I soon got it out of him. Millicent was two months pregnant, and threatening to expose him to his college authorities."

"If he didn't marry her?"

"Worse. If he didn't procure an abortion for her. He told me she'd been working on his nerves like a fiend. I could see it: he shuddered with disgust when he talked about her. But the practical problem seemed insoluble. Exposure would mean the end of his career, his vocation. Naturally, he would not entertain the idea of an abortion. Marriage, even if she could be brought to accept it, would mean tying himself for life to a woman whose nature he had plumbed deeply enough already. And besides, he couldn't begin to afford it financially: he was scraping through college on a legacy; our mother was then a widow, living on a small annuity."

Lighting a cigarette, Stephen puffed at it jerkily as he went on.

"Well, I hope I never have to live through anything like that April again. Peter was in a turmoil of panic, guilt and disgust. He'd fallen deeply in love with Millicent, only to discover that he'd fallen into a—a pit of—you don't know how a woman like her can flay a decent man. Peter was near suicide. She'd turned his love into an abomination, and then flung it back in his face—that's how he felt it."

"And that's how you came to write *Fire and Ash*?"

Stephen's head turned sharply toward Nigel. "Oh, you've read it? Yes. I started writing it then, to keep myself sane. Millicent never forgave me for that book—it penetrated even her thick hide."

Comprehension dawned on Nigel's face. "So she had a really strong motive for tampering with the proof copy of *Time to Fight*. I never thought your opposing the reprint of her novels was a strong enough one. She did it to get her own back for *Fire and Ash*. I suppose she thought you'd be bound to be held responsible for the stet marks. Or was she threatening to tell the partners that she had actually seen you doing it?"

"I've no notion what was in her mind," replied Stephen with some impatience. "There's no *proof* it was she who did the stetting, is there? If she did, you can dismiss any idea that she did it to expose Blair-Chatterley and avenge Paul, anyway."

"Why?"

"Because she never cared a damn what happened to Paul," Stephen bitterly replied. "As far as she was concerned, Paul never existed. That was part of the bargain."

"The bargain?"

"Yes. I went to see her. A fortnight after Peter had told me about it all. Thought I could plead with her, soften her somehow—I was nearly as ignorant and idealistic a young ass as Peter. I met her several times that month. Not at her home, of course; A.B.C. shops—that sort of place. It taught me a lot about women. She was only nineteen, but by God she knew all the tricks. Innocence betrayed, pathos, indignation, panic, evasiveness, the brave face—she tried everything. Well, I finally got her to realize that she could expect nothing more from Peter. I made her a proposition then—said I'd arrange for the child to be born up in the North, pay for everything, go up there with her, pretend to be her husband, and take the child off her hands. She made up some story for her parents that a London doctor had told her she'd got T.B. and a friend had offered to pay for treatment at a sanatorium. They seem to have accepted it. She was the most accomplished liar I've ever met, even at that age."

Stephen Protheroe fell silent. The brooding look on his face, in profile against the window, reminded Nigel of that time, over a week ago, when he and Stephen had talked about disinterestedness. "Perhaps we have lost what the word means," he had said. Well, Stephen had done one purely disinterested action in his life. He began to say this now, but Stephen cut in harshly,

"And a hell of a lot of use it was. It didn't take the load off Peter's conscience. He went into the missionary field and died of a tropical fever two years later in some God-forsaken place."

"And the boy? Paul? You looked after him?"

"Nobody else was going to." Stephen went off at a tangent. "Millicent's flair for evading responsibility was incredible. Of course, as I said, it was part of the bargain that I should take him off her hands. But she'd hardly given birth to him before she contrived somehow to dissociate herself from him entirely."

"Ah. The word Mrs. Blayne used."

"Who's she?"

Nigel explained.

"Did Millicent confide in her?"

"Not about the baby. But she told me that, even as a girl, Millicent could dissociate herself from her own mistakes, etc.—pretend they'd never happened, and take herself in quite successfully."

"Yes. I remember, when she was suckling the child—she had to do it for a bit till I found a wet nurse—she looked utterly detached from him, as though she was doing a favor to somebody else's baby. It was quite absurd, I couldn't help laughing once. She hated being laughed at. Humorless girl. When I told her why, she positively glared at me. 'I never wanted this creature,' she said; 'I never want to see or hear of it again, and I'm starting now.'"

"And did she?"

"Did she what?"

"Ever see or hear of Paul again?"

"No. She turned the whole episode into a bad dream. And I'd no desire to keep in touch with her."

"Not even when he was killed?"

"No. Paul was told his mother had died in childbirth. He accepted me as his father."

"But when Miss Miles turned up here—?"

"That was an exceedingly unpleasant reunion for us both."

"But surely you told her about Paul then?"

"Oh, I see, yes. The subject did come up presently. She had the infernal nerve to ask me how my son was. My son! I told her he'd been killed in action ten years ago."

"How did she take that?"

Stephen Protheroe grimaced, gave one of his volcanic sniffs.

"It was unbelievable. She went straight into a bereaved-mother routine. Why had I kept Paul from her? Why had I allowed him to get killed? I just saw red—went for her good and hard."

"That must have been the quarrel Jean heard. When Miss Miles called you 'Goggles.' Which reminds me that Susan—"

"You can understand why I didn't tell you all this before. Paul wasn't my secret alone."

Nigel rose to look at the photograph again. "There's a likeness to his mother."

"It was only physical, thank heaven."

"You became very fond of him?"

Stephen nodded. His face, from the noble brow to the small receding chin, looked suddenly smaller; the lips quivered for a moment. "After a bit I began to feel as if I was his father. He was a lovable little chap. Quiet and serious, like my brother, with plenty of fire behind it."

Stephen went on to tell Nigel how the baby Paul had been entrusted to foster parents—a Greengarth farmer and his wife, whom he himself had known from boyhood. He visited the farm from time to time, and later paid for the boy's education at a small public school. It was to support Paul that he had taken the job with Wenham & Geraldine in 1930, for his free-lance literary work was still not bringing in enough money. The farmer had offered later to adopt the boy; but Stephen felt it as a duty to his brother's memory that Paul should be "kept in the family." Paul had shown, as he grew older, an unusual aptitude with animals, and Stephen had been saving up to buy him a small farm when he came out of the army. But he never came out alive; so the money was spent, some of it, on the creature comforts with which Stephen was now surrounded. And a pretty poor substitute they must be, thought Nigel, for that promising nephew and all he had meant to Stephen.

"You realize," said Nigel, "that the police may inquire into all this? Is there anyone who can corroborate your story?"

The farmer and his wife were still alive. To protect his brother, Stephen had told them he was the child's father and that its mother had left him shortly after the baby was born. How far they had believed this fiction he never knew; but they agreed to tell Paul, when he asked, that his mother was dead. Nigel was welcome to their name and address. Other confirmation there could be none, for the whole affair had been conducted with the utmost possible secrecy. Stephen and Millicent had lived together in rooms at Greengarth for several months before the child was born, as man and wife. In view of Millicent's condition, they occupied separate bedrooms. It had been a protracted and grisly farce, keeping up the pretense.

"But you see," Stephen concluded, "you've only my word for much of this. I'm quite certain Peter never confided in anyone else."

"Was he alone when he died?"

Stephen looked puzzled for a moment. Then he said, "Oh, a deathbed confession? I see. Well, the C.E.M.S. might be able to help there. But I doubt it. By the time they'd got Peter back into the hospital, he was in a coma, and never spoke again. His Bishop wrote and told us that, when they sent home Peter's belongings."

Nigel took down the few names which Stephen could provide, but not hopefully. The threads of this story, as of so much else in the case, went too far back into the past, and the most important ones were broken—Millicent and her parents were dead, Peter Protheroe was dead, Paul was dead.

Nor was Nigel disposed to doubt Stephen's account of his relationship with Millicent. Unfortunately this account gave no solid ground for a further step in the investigation of the *Time to Fight* problem. However fond of his nephew Stephen had grown, it was not very likely that he should have tampered with the proof copy, nearly ten years after Paul's death, to expose the Governor whose ineptitude had caused it. Yet it seemed even more improbable now that Miss Miles had been the culprit. She had wished never to see or hear of Paul again.

"There are some people," said Stephen, as if he had read Nigel's thoughts, "who create the maximum nuisance, in any way that lies to hand, simply because they *must* assert themselves. They are the truly irresponsible ones."

"You're thinking of Millicent Miles?"

"Yes. And the same goes for Cyprian Gleed, I'd say."

"Both of them vindictive characters."

Stephen considered the word. "Malicious, perhaps, rather than vindictive."

"Millicent Miles hated you," Nigel reflectively continued. "You knew of a shady episode in her past. You told her exactly what you thought of her. You opposed the reissuing of her novels. She could have stetted those libelous passages, on an impulse, with a vague idea of getting you into trouble—and, of course, getting her own back on the author for ragging her in the mess. Several smallish motives adding up to quite a big one. Yes, it's in character."

"But you don't feel confident about it?"

Nigel soon rose to go. If Stephen was the big fish for whom he was angling—and it seemed wildly improbable now—the bait had not been taken. Not even a nibble.

XIV *WRONG FONT*

"I'm afraid this isn't a very convenient time, my dear fellow. I'm up to my neck in—" Arthur Geraldine gestured toward the neat piles of papers on his desk. The senior partner's tone was pleasant enough, but his gray eyes had a chill on them.

"I'm sorry," said Nigel, "but there's something I must clear up. It really can't wait any longer."

Geraldine sat back, stretching out his arms in front of him, the strong, hairy hands, clenched into fists, resting on the desk.

"Very well then. What is it?"

"Why did you tell me you'd never met Miss Miles till your firm took her over?"

Arthur Geraldine's eyes went chillier still. "I don't like your tone, Strangeways. May I remind you that you are, temporarily, an employee of my firm?"

Ignoring the frigid rebuke, Nigel went on, "Miss Miles told me that she had met you many years ago, under rather peculiar circumstances." He refrained from mentioning her remark that Geraldine "always was a weak man."

"I don't recollect it. But it's perfectly possible. I've met a good many people in my life."

"Isn't it rather odd that she never reminded you of this meeting?"

Geraldine shrugged; but there was a watchful look on his face now.

"Didn't she remind you of it?" Nigel persisted.

Geraldine's long, thin mouth stretched tighter. "Are you suggesting I'm a liar?"

"There's no use stone-walling with me, Mr. Geraldine. I have evidence that, when you were a representative of the *Daily Sun*, you bought a Rockingham dinner service from Miss Miles' parents in Wimblesham."

A flush spread slowly over the large brow and bald head. "Yes, I did acquire the Rockingham set there, I remember. But I'd no idea it was her parents who—"

The door opened, and Stephen Protheroe trotted in. "You left this behind," he said, handing Nigel a bottle of nose drops.

"Oh, many thanks." Nigel absently set down the bottle on Arthur Geraldine's desk, as Stephen went out again. "Did you actually meet Miss Miles when you visited her parents' house? She was a schoolgirl then."

"Ah now, how could I possibly remember that? It was over thirty years ago." Geraldine almost crooned it. He had the Irishman's gift for fluent changes of front; he was all affability and helpfulness now.

"Miss Miles remembered it all right. Well enough to mention in her autobiography a man she met at that period and gives the pseudonym of Rockingham." Nigel slid smoothly off this very thin ice. "Did you go to the house often?"

"Twice at least, I think. It's beginning to come back, you know. I do seem to remember a girl there—a toothy, intense creature—do you tell me that was Millicent Miles? Well, bless my soul!" Arthur Geraldine grinned like an amiable shark. "Yes, she was mad keen for her dad to take out a subscription to the paper, so she'd get a set of novels I was hawking round —one of those newspaper free-gift schemes to increase circulation."

"And did he?"

"No. I think he must have been on the rocks. Anyway, he told me to call again. The second time, his wife wasn't there. He asked me what I'd give him for the Rockingham set—I'd admired it on my first visit. I made him an offer, and he jumped at it. But I had to take the service away at once— before his wife returned, I suppose. I remember emptying my suitcase for it. I was carrying round some specimen sets of the novels. I said I'd come back for them the next day. There was a blinding row when I did. Mrs. Miles went for me like a fury. But I had her husband's receipt for the money, so there was nothing she could do about it."

Arthur Geraldine's expression was almost ingratiating, as he added, "It's terribly shocking to look back on, I agree. I fleeced the poor fellow. Collectors have no morality at all, you know. It's like a physical passion. You fall in love with some object, and you've got to have it. Besides, I hadn't much money myself those days."

Nigel meditated this remarkable statement. It was not the confession itself which was remarkable, but the fact that Geraldine should so gratuitously have made it. "And Miss Miles never brought this up, you say?"

"Never breathed it, I do assure you." Arthur Geraldine was positively beaming. "I see what's in your mind, my dear fellow. Blackmail, eh? It was a discreditable episode in my life, I'm not denying it. You can see why I didn't want it to come out. But I'd not kill the poor woman in order to keep it hushed up."

Nigel agreed that this was, on the face of it, most unlikely. But why, he thought, have I been able to extract the rotten old stump so painlessly? Geraldine showed no resistance, never challenged my sources of information. Is he flourishing before my eyes that deplorable episode of the Rockingham set, simply to conceal the existence of something far nastier behind it?

"You gave Miss Miles very favored treatment here," Nigel said to gain time. "I'd not have thought her a Wenham & Geraldine kind of author, anyway."

"Every publisher has to compromise between his literary

standards and his pocket. Besides, Basil Ryle was keen; he rather bounced us into taking her on."

"A singularly unfortunate decision," Nigel remarked dryly, "considering that she involved you in a libel action and then got murdered on the premises."

"What's that? You mean it was *she* who tampered with the proof copy? In God's name, why?"

"General mischievousness and particular malice. I know for a fact that she had private grudges against General Thoresby and Stephen Protheroe. It'd round it off nicely if she had a grudge against you, or Miss Wenham."

"I don't follow you."

"If, for example, she'd tried to blackmail you over the Rockingham set, what would you have done?"

"Told her to go to hell, of course. But—"

"Which would have given her a motive for damaging you through your firm."

"But I keep on telling you, she never referred to the blasted Rockingham, let alone—"

"Let alone anything else that happened when you visited her home in 1924?"

Geraldine's bald head flushed pink again. It could have been, of course, through the effort he was making to repress indignation and anger.

"Will you please make yourself plain?" he said, in a constricted, dangerous tone. "I don't like innuendoes."

"The trouble with Miss Miles is that no one ever knew whether she was telling the truth or romancing."

"I asked you—"

"Yes. She confided in a school friend that a man who came to her house, in the summer of 1924—a man with very thin lips—had attempted a criminal assault upon her. As a blackmailing card, a criminal assault on a schoolgirl, if there was other confirmation of it, would still be fairly potent, even after thirty years."

Nigel started, as the ruler which Geraldine was playing with snapped in half. "You're talking great nonsense," said Geraldine in a muffled voice. Nigel waited for more; but the senior partner, with a preoccupied look, took up a sheaf of papers from his desk. "I really must get on with my work now."

Going upstairs, Nigel sought out Stephen Protheroe again. "Sorry to keep pestering you. It's just this. Mr. Miles went bankrupt in 1924. You met Millicent first in 1925, when she was working in the bookshop. Did she ever invite you to her house?—she was still living at home then, I believe."

"Yes, I went there once or twice."

"They were very hard up still, presumably?"

"Not that you'd notice, no. I seem to remember getting the impression that Miles had fallen on his feet somehow."

"Miss Miles never discussed his financial affairs with you?"

"I don't think so. Apart from a little self-dramatization as the struggling young genius who had to support her ruined parents."

"Which she could hardly do on her wages as a bookshop assistant."

"No. It was rather odd, I suppose. Her father had been sacked by his firm, and I'm pretty sure he hadn't found another job when I first met her. But I really knew very little about that side of her life. We spent most of our time discussing Culture—or rather, Millicent's contribution to it. Young men are always flattered to have female disciples."

Arthur Geraldine's secretary came in. "You left this behind in Mr. Geraldine's room, didn't you?" she said, handing him the bottle of nose drops.

"Oh, thank you very much."

"You seem to be always leaving it about," said Stephen. "Put it in your pocket, for heaven's sake."

Nigel now told him that he wanted a few words with Susan Jones. Stephen rang the head of the invoice department, who said he would send her up in a few minutes.

"You want to talk to her privately?"

"Yes."

"Well, you'd better use the next room."

"I don't think that'd be quite—"

"Good heavens, man, it's not haunted, is it? They've cleaned it up," said Stephen irritably.

"Nevertheless, Susan might be a bit squeamish. Could you ask Jean to clear out of the Reference Library for a bit?"

"Oh, I suppose so. Heaven knows what she does there all day, anyway."

So Nigel's second meeting with the blonde bombshell took place at the venue of his first. Susan was nervous, though, this time, and far less forthcoming.

"Do sit down, Susan. Jean said you wanted to see me about something. Sorry I've been so long getting round to it."

"I'm sure I don't know—" the girl began, tossing her head uneasily.

"Trouble with the boy friend? Just give the word, and I'll challenge him to a duel."

Susan smiled wanly. "You're in with the police, aren't you?"

"In trouble with them. Constantly. Why?"

"I've never had anything to do with that lot. You can't trust them, can you? Beating up prisoners and taking bribes and that. But it's really what my boy friend would say, see? He's refined."

"The one who reads books?"

"Who else?"

"Well then, what would he say about what?"

"Besides, Miss Wenham doesn't like it. She's ever so fierce on—"

"Miss Wenham. How does she come into this?"

"I'm telling you, aren't I? She doesn't like we girls going into the packers' room. But what I say is a girl wants to see a bit of life before she settles down."

"And the packers' room is the place to see it?"

Susan gave him an unexpectedly gamine glance. "You're

telling me! Mind you, I wasn't doing any harm. But it might be mis—mis—"

"Misinterpreted?"

"Look what education does for you! You know, you remind me sometimes of David. That's my boy friend. My steady, I mean."

Nigel's almost limitless patience was nearing exhaustion. But, after several more diversions, he got Susan back onto the main road, to find himself amply rewarded. Last Friday night, she told him, she had a date with "one of the gentlemen in the packers' room." She left the Reference Library sharp at 5:30, went down in the lift, and entered the packers' room. All the other packers had departed by now, and her "date" took the opportunity of the room being empty to embark on some preliminary dalliance—this latter being Nigel's interpretation of Susan's coy statement that "we chatted awhile before going out." After about ten minutes, they turned off the one light still burning there, and opening the door Susan peeped cautiously out—cautiously, because they had just heard the lift gates closing, and because she had no business to be there, Miss Wenham disapproving of this sort of fraternization on the premises.

Susan peered out, then, into the passage which led past the lift, the foot of the stairs and the reception-room door to the side door of the building. The passage was lit at that time of the evening, as Nigel knew, by one low-powered bulb. The light was sufficient, however, to show Susan a man who was just going through the inner swing doors at the end of the passage toward the street door. She had only seen his back, and that for a moment only, but her description of him was clear enough as far as it went.

"It was a fairly small man," she said. "He had one of those duffel coats on like the Teddy boys have started wearing. Biscuit-colored, I think. And a dark hat with a wide brim. And he'd got sort of hair at the side of his face—he was turning just a little to get his bag through the swing door."

"What kind of a bag?"

"Well, you know, a grip bag. Must have been fair-sized and heavy, because he grunted a little getting through the door with it."

"Did Whatsisname—your packer friend—see this chap?"

"Oh no. There wasn't room only for me to look through the door."

"What did he think about it?"

"George? I didn't mention it to him. I never thought anything of it at the time. I mean, I thought it was just someone who'd been visiting one of the partners."

"And there was nothing familiar about this chap?"

"Ooh, he hadn't time to be familiar—I've told you—he was beetling out of the office."

"I mean, you didn't recognize him? His back view didn't remind you of anyone you know?"

"N-no." The girl's negative did not sound absolutely firm, but Nigel pressed her no further.

"What does George say about all this now?"

"Oh, I've not mentioned it to *him*. It's none of his business, is it?"

"In fact, you've not mentioned it to anyone?"

Susan, looking a little shamefaced, jerked one shoulder forward with a petulant movement. "Well, I've told you, haven't I?"

"But the police. They asked if anyone here had seen anything out-of-the-way last Friday evening."

"There's nothing out-of-the-way about a man leaving the building."

"Oh my dear Susan, why were you worried about it, then? Jean saw you had something on your mind."

"Now don't you start nattering at me! I've had enough worry— It's what I told you. I don't want my boy friend to hear about it, see? Or Miss Wenham."

"But you'll have to tell the police your story."

"It *isn't* a story!"

"Your account of what you saw."

"I'm not having anything to do with that lot. We're respectable people. My dad would tan me if I got mixed up with the police."

"Did you ever see Cyprian Gleed, Miss Miles' son, when he visited the office?"

"Not that I know of. What's he to do with it?"

Nigel rose to his feet, and Susan did the same.

"Look, Susan. Inspector Wright has got to be told about this. I can tell him myself; but it'd come far better from you. Why not get it over with at once?"

"He'll half murder me for—"

"He'll do nothing of the sort. The Inspector is a very nice man and a friend of mine. I'll ring him up to fix an interview. And I'll hold your hand while you talk to him."

Susan's expression showed that she did not take the last phrase figuratively. The languishing look was soon followed, however, by one of alarm.

"But will I have to give evidence in court? I reelly couldn't."

"You may not have to. It all depends. But if you do, you'll be the star witness for the Crown."

The word "star" evidently got home. Susan's huge blue eyes brightened. She saw herself in the witness box, switching on all her charm. And then there'd be the newspapers.

"Give my boy friend something worth reading about, wouldn't it?" she said with a giggle, taking up the dislocated stance favored by modern fashion models.

"You'll knock them cold," said Nigel, and reached for the telephone.

"I'd like to have put that blonde across my knees and given her a good hiding," declared Inspector Wright. "Here we've been, wasting a dozen men's time for a week, just because she chose to lock up this information in her silly little head."

It was nine o'clock on the same evening. Nigel had dined with Clare Massinger in her studio, and she was now making another pot of coffee for the Inspector, who had just dropped in. His long, sallow face looked gaunter than ever after the intensive labors of the last week, and the brightness in his eyes seemed almost feverish.

Clare poured him out a large cup of coffee and tipped some brandy into it, then curled herself up close to Nigel on the settee. Wright raised his cup to her: "Your health, Miss Massinger. This'll save my life." He glanced round the studio, nodded at an exhibit, and with one of his dazzlingly vivid pantomimic gestures sketched the movement of a sculptor's hands molding the clay. "How's the old slap-and-tickle going?"

"Mustn't grumble."

Clare and the Inspector enjoyed keeping up the fiction that he was a deplorably uncultivated person whose idea of art went no further than Highland cattle and indecent postcards. He teased her for a little on these lines, accepted another laced coffee, fished out a ball of clay from between the seat and the side of his armchair.

"Funny places your guests park their chewing gum," he observed.

"I'd like to do your hands one day," said Clare.

"What's wrong with my face?"

"Oh, that's quite interesting too. But you'd never sit still long enough for a portrait bust."

"I'll sit still for a month when I've got this case tied up."

"But I thought it was."

"Bless your heart, lady, it's only just begun. Chap answering to the description of Cyprian Gleed seen leaving the premises just after the time the murder could have been committed. All right, we pull him in for an identification parade tomorrow—set him in line with half a dozen other types wearing duffel coats and large black hats."

"And beards," said Nigel.

"All right. We can rustle up false beards for them. But do

you think that nitwit Susan Jones will be able to pick out our man? It'd be a miracle. And they don't happen—not to me."

"So you're sure he's your man?" asked Nigel. "What about that page inserted in the autobiography?"

"Lord love you, Mr. Strangeways, that's the strongest bit of evidence against him, to my mind. Not that the D.P.P. would look at it. You've been barking up the wrong tree. The whole thing's quite simple. Young Gleed is desperate for money, and he hates his mother. When I first interviewed him, he pretended not to know that she'd made no will. But he might well have known that she hadn't. All right. She's going to die intestate. Now he's been in and out of her room at Wenham & Geraldine's: he's had plenty of opportunity to read her book. There's a page in it which states or suggests that she'd had a child by another man before her marriage to Gleed's father. This page may not have made it clear she never married the man. That doesn't greatly matter, because in his last conversation with you young Gleed showed he knew nothing about the law of intestacy—didn't know that a bastard has no standing at all where there is a legitimate child. The page he typed after cutting his mother's throat was inserted to ensure *that the authorities should never look further than himself for the next-of-kin.*"

"Yes," said Nigel slowly, "that seems plausible enough. But aren't you forgetting the other thing the murderer did to the typescript?"

"I'm forgetting nothing," Wright strenuously replied. "You're thinking of the erased capital G further back? Look—imagine you're Cyprian Gleed. You've killed your mother, you've taken out the give-away sheet and replaced it with the fake one you've just typed. You riffle back through a few pages of the typescript, and suddenly your own initial stares you in the face."

"But—"

"In that state of mind, he could easily mistake a G for a C. Cyprian. He's in a hell of a hurry to get away—no time to study the reference of that capital letter. It just rises up off the page and points at him, like his dead mother's finger. So he hurriedly rubs it out."

"You're an awfully imaginative man, Inspector," said Clare, shivering a little.

"Yes," said Nigel. "I hadn't thought of that possibility."

Inspector Wright now outlined his next moves. Cyprian Gleed was under close surveillance. Wright had interviewed him again this afternoon, but without bringing up Susan Jones' statement. Gleed had denied buying recently either a duffel coat or a grip bag. Armed with photographs of the young man, Wright's team would intensify their efforts to trace such purchases and that of a cut-throat razor. The appeal to the public had so far produced nothing useful; but, now that Susan Jones' statement had narrowed down the time limit, a further appeal would appear in tomorrow's newspapers, based on the probability that the murderer had thrown his bag off Hungerford Bridge at about 5:45, five minutes after leaving the publishers' office. It was already established that the duffel-coated figure had not been a bona-fide visitor.

"What puzzles me," put in Nigel at this point, "is that he should still have been wearing the coat when Susan saw him. It must have been bloodstained. Admittedly, the staff have all left by 5:40 P.M. on Fridays; but there was still a risk of bumping into one of the partners. Why didn't he take off the duffel coat and put it in the bag before leaving Miss Miles' room?"

"Probably just forgot to. Nerve snapped at last. Anything to get out of that room double-quick."

"Do you suggest he walked out into the street covered with blood?"

"He could have taken off the coat and stowed it away after he'd gone through the swing door and before opening the street door." Inspector Wright's burning eyes were fixed upon Nigel's. "Never satisfied, are you, till you've crossed all the t's."

Nigel shrugged. "I expect you're right. But I'm still bothered about that duffel coat. He did everything else so very much according to plan. And there was something Susan said to me—or rather, the way she put it. You gave a sort of unconscious echo of it, yourself, a few minutes ago." He broke off, looking abstracted.

"The maestro's got an attack of mystification," Clare remarked.

"There are times when a uniform can be the best disguise. Remember Chesterton's story about the postman? But this would work the opposite way—a murderer *wanting* himself to be seen. Ah well." Nigel turned suddenly upon the Inspector. "Have you considered Arthur Geraldine in the role of a dark horse?"

"Now where are we off to?"

Nigel described his interview with the senior partner after lunch. Suppose that Millicent Miles' original story, as told to Mrs. Blayne, *had* been true. Suppose the young Geraldine did try to assault that extremely mature schoolgirl. The "blinding row" which took place, Geraldine had said, over the sale of the Rockingham set might have been in fact over his attempt upon Millicent. Her father might have agreed to let the matter go no further, in return for cash. Certainly, a year later, though he had not got another job, he had seemed to Stephen to be living quite comfortably. Mr. Miles could have no hold over the young Geraldine without a written confession by the latter. He would hush the matter up, in return for an annual payment, say, and the confession as security for it. When Mr. Miles died, a few years after his wife, this confession would pass into the hands of his daughter. At that period she was affluent. But recently, with her books out of fashion, might she not have brought the document into circulation again?

"It's a preposterous idea," said Wright warmly. "A young chap might fall for that sort of intimidation; but Geraldine isn't a callow youth now; he'd go straight to the police, if Miss Miles threatened him; he'd know that the plaintiff's name isn't mentioned nowadays in blackmail trials of this kind."

"That's all very well in theory. But in practice he couldn't be sure that his name wouldn't come out. And a respectable publisher's reputation couldn't survive that sort of scandal—not when it was backed by a written confession."

"All right," said the Inspector in a humoring way. "So he murdered her to stop her mouth. Where's the document, then—the confession? It's definitely not among her papers. But if she brought it to the office last Friday, to bargain with him, don't tell me he couldn't have taken it from her without having to slit her throat first. Or alternatively, he'd buy it from her, in which case the murder would be pointless."

Nigel made a defeated gesture. The sinus pain was bad this evening, blunting his intelligence. He took from his pocket a bottle of nose drops, filled the syringe and unscrewed it, placing the bottle on the floor beside him. Clare got up from the settee to make room for Nigel to lie full length on it and hang his head backward over the edge. In doing so, her foot caught the bottle, knocking it over.

"Oh, damn it all, Nigel, why will you always leave things on the floor?"

"Sorry, darling."

"It's gone all over the rug." Clare looked about for an old cloth, then bent down to mop up the liquid. "*Nigel!*"

"What's the matter?" Holding the filled syringe, Nigel sat upright. Clare, rigid, pointed at the stain on the rug, from which a thin, vicious thread of smoke was now rising. The next instant she stamped her foot on the stain, as if it were a viper—stamped again and again, till Inspector Wright pulled her away from it. The black rug, where the liquid had splashed out, was turning a dirty brown. Bright-eyed, the Inspector regarded it.

"Ha! Oil of vitriol."

"What?" Clare was glaring stupidly at the bottle. She stooped as if to pick it up.

"No! Don't touch it!" said Wright sharply. "Mr. Strangeways, give me that syringe."

Clare whirled round, her face dead white.

"Oh, Nigel! Did you—?"

"It's all right, love." Nigel held up the syringe, still full, for her to see, then handed it to Wright.

"I'm always knocking things over," moaned Clare, quite dazed.

"Lucky for Mr. Strangeways you are."

Clare's rigid face softened. Crumpling, she fell forward into Nigel's arms, gazed up at him incredulously, and stroked her fingers along his cheek as if to reassure herself that he was still there.

"What would have happened?" Her voice was shaking, almost inaudible.

"Well, a syringeful of vitriol, injected into the sinus—"

Clare laid her trembling hand on Nigel's mouth, and burst into tears.

When she was calmer, Inspector Wright asked the inevitable question.

"I don't know," said Nigel. "Protheroe and Geraldine both had the bottle in their possession today: I left it behind in— no, what a fool I am! This thing has addled my wits. I finished that bottle, and took a fresh one out of my medicine cupboard before coming along this evening."

Clare shuddered convulsively in his arms.

"And who had access to your medicine cupboard?" asked Wright.

"Well, it's in the loo." Nigel frowned. "The other evening, when Cyprian Gleed came to see me, he asked if he might use it. My goodness, and he—"

Nigel stopped abruptly. He did not want to harrow Clare's feelings any more, so he refrained from repeating Gleed's last words to him on that occasion—

"I'll let myself out, when I've finished. *You probably won't be seeing me again.*"

XV *CLOSE UP*

"**B**ut why should he do it? Such a foul, vicious thing? It's what you'd expect of a—a corrupt child."

"He is a corrupt child. It was an act of pure spite. I'd wounded his vanity pretty deeply once or twice before that evening he came to my house. And I waded into him unmercifully then. I must say, I hardly blame him—"

"Oh nonsense, my darling. Don't be so bloody Christian and understanding. I'd like to toast him over a slow fire."

Clare and Nigel were breakfasting the next morning. There were shadows under Clare's eyes, and her long black hair had an enchanting glossiness. She smiled dreamily at Nigel, then seemed to withdraw into herself, with that quality of elusiveness which had captivated him the first time they met.

"I hope you're going to keep up this habit of saving my life. That's twice so far."

Leaning over, Clare laid her mouth on his and held it there for a long time.

"Do you know what you are? A cross between a black cat and a white camellia," said Nigel.

"You make me tremble every time you come near me. Still."

"That's a good thing isn't it?"

"A very good thing." Clare rose, and sat down on the other side of the table. "I feel like a convalescent. All weak and wondering. Look, the sun's come out."

"Busy old fool."

"What'll happen to him?"

"Oh, Wright'll trace where he got the stuff and pull him in for attempting grievous bodily harm or whatever they call it."

Clare Massinger took a lump of clay and began kneading it with her small, strong fingers.

"Irresponsibility," Nigel continued. "That's Gleed's trouble. The inability or the refusal to envisage the full consequences of an act. His mother was the same; but she had talent of a sort, to keep her on a more or less level keel."

"I knew a child once who strung a piece of wire across a drive to decapitate her governess bicycling back in the dark."

"I hope you were soundly thrashed for it."

Clare colored deliciously. "I never said— Oh, well, yes, I was. Luckily the wire had sagged, through my own incompetence, by the time she got back, so it only caught her front wheel. But my parents took the will for the deed, as they say."

"Whither is this gruesome reminiscence tending?"

"The point is, I'd not have stuck a knife into that governess, or pushed her off a cliff. Yet I was prepared to attempt long-range decapitation. Children have no imagination—I mean, I visualized with great relish the wire whipping off her head; but in a way I didn't believe it could happen—it was like an experiment that someone else would be bound to stop before it got too dangerous. Children's imagination is so limited."

Nigel watched her attentively. The lump of clay, which she absently molded, was taking the form of a rudimentary horse. "A deficient sense of reality. Yes?"

"It does seem odd to me," said Clare in her light, high voice, "that a chap who could fix up that long-range, delayed-action thing for you could also walk up to his mother and cut her throat. I'd have expected him to use poison on her, if anything. But perhaps we've analyzed him all wrong."

"Perhaps," said Nigel, giving her a noncommittal look.

When he got back to his own house, half an hour later, Nigel found a telephone message from Liz Wenham, asking him to go and see her as soon as possible: she would be at home all the morning.

The room in which Miss Wenham received him was lined with books and filled with knickknacks. She herself, dumpy in a tweed suit of a rather unbecoming pattern, clear-eyed, decisive in manner, resembled more than ever a Victorian blue-stocking or a member of some North Country family with a tradition of sweetness and light—more light than sweetness, perhaps—behind it: one of the great intellectual dynasties which have intermarried and preserved in the raffish modern world an earnestness, a faint academic aroma, a confident moral tone and a general air of unremitting mental hygiene, inherited from their vigorous forebears.

She herself was sitting behind a ponderous reading stand which, Nigel was at once convinced, had belonged to the famous James Wenham. Indeed, if Cyprian Gleed's room was a mausoleum of false starts, this one was a museum of successful finishes: the furniture, the photographs, the ornaments, above all the books (autographed, no doubt, by their celebrated authors, with obliging tributes to the Wenham dynasty), conveyed the impression of a solidity too grand to be labeled as smugness.

"Well, Mr. Strangeways, when is this mess going to be cleared up?" was Liz Wenham's crisp opening.

"In a few days, I think."

"The police are satisfied they know the identity of the murderer?"

"Ye-es, I believe they are."

"But you are not satisfied?"

"It's not my department. I'm only concerned with the libel trouble."

Liz Wenham's gray eyes sparkled frostily. "Come, come, Mr. Strangeways, don't let us be evasive with each other."

"Well then. Cyprian Gleed had motive and opportunity. Moreover, he—or someone closely resembling him—was seen leaving the office at 5:40 on the night of the murder."

"This is news to me," said Liz quickly. A certain fullness in her voice suggested to Nigel that it was good news. He gave her an edited account of Susan's statement.

"I shall have to speak to that girl. What was she doing in the packers' room? Cuddling, I suppose." Cuddling might have been equivalent to Babylonian orgies, the way Miss Wenham made it sound.

"It's called 'necking' nowadays," Nigel could not refrain from saying.

"Really? What an *inadequate* word! But I didn't ask you here to discuss erotic synonyms." Liz Wenham was looking at him very straightly. "You'll have realized by now that the firm of Wenham & Geraldine means everything to me. I would stop at nothing to preserve its good name." Liz Wenham emphasized her remarks by beating her fist on her knee; and there was indeed an almost fanatical look in the gray eyes.

"Ah," said Nigel, smiling pleasantly at her, "you're going to ask me to suppress evidence."

Liz flushed deeply, but answered without resentment: "Not about the—the death of Miss Miles." She clenched her fist again. "Arthur Geraldine made an extraordinary confession to me last night. I have his permission to pass on the gist of it to you. He did not feel disposed"—Liz looked momentarily uncomfortable—"to face you directly with it. I understand you had already broached the matter to him."

"About his attempted rape of Millicent Miles when she was a schoolgirl?"

Liz Wenham flinched at this plain speaking, but continued. "Your facts were mainly correct, but your interpretation is wrong. Arthur assures me that it was a put-up job between the girl and her father, though he did not perceive this at the time. The girl—er—maneuvered him into a compromising situation, whereupon the father entered the room and surprised them."

"Well, that's certainly a new angle on the story."

"Don't you believe it? It has happened often enough,

though not perhaps with a schoolgirl as the agent provocateur. Arthur was young and inexperienced. He's not got a very great deal of backbone, you know. Anyway, he was made to sign a confession, and he paid the father regular hush money for several years till Mr. Miles died."

"And then the daughter took over?"

"No. There's no question of that." According to his own account of it, Liz Wenham continued, Arthur Geraldine had not been blackmailed by Miss Miles. He had given her favored treatment, when the firm took her over, partly because she might prove a remunerative author herself and also help the firm through her contacts in the literary world, and partly, as he had admitted to Liz, from anxiety lest she should begin raking up the past. But in fact she had never referred to it, or to the written "confession" her father had extracted from him, and he assumed now that it had been destroyed after Mr. Miles' death.

"So there you are," Liz concluded, with a challenging glance at Nigel. "It's a deplorable story, but nothing worse. And I rely on you to keep it to yourself."

"Inspector Wright and I would certainly not publicize it. But we could not prevent it coming out in court, if Mr. Geraldine was charged—"

"But you *must* see," Liz impatiently exclaimed, "he had no possible motive."

"If the account he gave you is correct. We've only his word for it. Or rather, yours."

"Are you suggesting that *I* made it up?"

"You said just now you'd stop at nothing to preserve the good name of your firm. Wait a minute! The account you've given sounds convincing enough to me. But it cannot clear your partner of all suspicion."

"But Cyprian Gleed was seen—"

"Someone wearing a duffel coat, a dark hat and a beard was seen. A smallish person. Neither Mr. Geraldine, nor Ryle, nor Protheroe is tall. Nor, for that matter, are you. The footprints in Miss Miles' blood show that the murderer was wearing size 10 galoshes. None of you has large feet: you could all get that size of galoshes over your shoes."

Nigel was watching Liz Wenham closely while he spoke. At one point, he noticed, her whole expression changed, making her look vulnerable, almost helpless.

"But Stephen left the office well before 5:40. I thought that was established," she said. It sounded less like a statement than an appeal.

"Oh yes, he did," Nigel drew out a long silence. "What a charming flat he has. I don't wonder he seldom leaves it."

Liz seemed quite at a loss with this change of subject.

"Do you know the people he was staying with last weekend?" asked Nigel.

"Yes. The husband is one of our authors."

"A close friend of Mr. Protheroe's?"

"I couldn't say how close. But I believe Stephen had a standing invitation to go down there."

"Of which he didn't avail himself till last weekend?"

"Mr. Strangeways! I dislike insinuation, and I dislike tittle-tattle." Liz Wenham was suddenly very angry indeed. Her dumpy body went quite rigid. "It would be more straightforward to ask Mr. Protheroe these questions."

No, thought Nigel, the firm of Wenham & Geraldine doesn't mean *everything* to you. He said, "I'm sorry. I have a devious mind. And in a murder investigation the straightforward route is not always the most profitable one."

"But you are only concerned with the libel trouble. You said so yourself."

"Now it's you who are being evasive, I think. Besides, the libel may be linked with the murder." Nigel paused again. "Has Stephen Protheroe ever talked to you about his brother, Peter?"

"The missionary? No. He died years ago, didn't he? Stephen has always been reticent about his private life."

"So you've no idea what produced *Fire and Ash?*"

"A poet produced it," replied Miss Wenham, in her forthright manner.

"But not out of thin air. No one, the age he was then, could just *imagine* so much naked experience."

"He has a great capacity for suffering, no doubt."

"But not for writing out of it, any longer. Tell me," Nigel went on, "how is Basil Ryle nowadays?"

"He seems to be coping a bit better," Liz coolly replied. "You had a talk with him, didn't you?"

"Yes. He was in rather a taking—had the idea he might have killed Miss Miles in a brainstorm."

"I'm afraid he's not a very well-balanced character. The war, you know." Liz Wenham's eye wandered, and she spoke in a distrait manner. "Not that he isn't brilliant at his work. It was something of a gamble when we took him into partnership, but it has certainly been justified. A great pity he got himself mixed up with that woman. A brainstorm, did you say?"

"Yes. So he feared."

Liz Wenham gave signs of wishing the interview to be prolonged no further—such clear signs of impatience that Nigel, having offered to let himself out, paused for a moment outside the drawing-room door, and heard Miss Wenham dial a number and presently say, "Stephen? It's Liz. Could you lunch with me? At home?" Her voice, caught up in some extreme urgency of emotion, was barely recognizable. She is a very intelligent woman, thought Nigel as he walked away down the street, and yet somehow a very guileless one.

Late that same afternoon, Inspector Wright rang Nigel at home. The result of the identification parade had been negative, he said: Susan thought she recognized Cyprian Gleed as the person she had seen leaving the office, but was not prepared to swear to it in court. On the other hand, the source of the vitriol had been traced—a friend of Cyprian Gleed's who worked in a manufacturing chemist's, had supplied him with it. Faced with this evidence, Cyprian had "shammed dead," as Wright put it—refused to make any statement without legal advice, and been duly charged and arrested. The appeal to the public had so far produced no eye-witness of the hypothetical doings on Hungerford Bridge, thus confirming Inspector Wright in his view that Londoners were an unobservant race. The police had failed to trace the recent purchase of a grip bag, a duffel coat or a razor by Cyprian Gleed.

"You're wasting your time on that, surely. He'd never walk into Wenham & Geraldine's in his usual attire, if he was going to commit murder. There'd be some attempt at disguise," said Nigel.

"We're inquiring about the purchase of a light macintosh, too, of course."

"A light macintosh?" Nigel was taken aback.

"Yes. One that'd roll up fairly small and could be stowed easily in the grip bag. A large-size macintosh."

"Oh, I see. You mean, he'd wear it over his duffel coat, to keep the blood off it, and pack it away as soon as he'd done the job?"

"Yes. He has an old mac, but there are no traces of blood on it. Therefore we look for a new one."

"You think of everything."

"I'm paid to, Mr. Strangeways."

"But have you thought about the hat? A large, black hat? Very few people wear them in this country. What a give-away!"

"Maybe he couldn't afford a new hat, as well as a mac. He was broke—"

"Oh bosh! Anyone can steal a hat."

"But he knew the office would be almost empty after 5:30. He could be pretty certain there'd be nobody on the top floor or the ground floor, at any rate, and the lift took him past the floor where the partners work."

"And the key of the side door?"

"Ah, if we could trace that to him, it'd be a push-over. I've had yet another go at Miriam Sanders—she's disillusioned at last about young Gleed, by the way, after being told what he tried to do to you—but she swears she never lent him the key, and I believe her. He knew where it was kept, though, and he was in the office the day it disappeared. I've no doubt he distracted her attention for a moment and pinched it."

Sunday passed without event. At half past ten on Monday morning, while Nigel Strangeways was digesting his breakfast and a report from the Greengarth police, which Wright had sent him by special messenger, the telephone bell rang. It was Arthur Geraldine. In a subdued tone, as though he were condoling with Nigel over a bereavement, he announced that a most unfortunate occurrence had just taken place in the office. Could Nigel spare the time to come along immediately?

"What is it? What's happened?" asked Nigel, half expecting to be told of another death on the premises.

"The—er—weapon has come to light. A razor. It puts me in a very awkward position. That's why I rang you first."

Awkward position, indeed! thought Nigel, fuming with irritation as the taxi bore him eastward. I bet I know where the razor was found.

He would have won the bet. Five minutes before Geraldine had telephoned, his secretary entered Basil Ryle's room.

"Mr. Geraldine would like to look through the *Aston Memoirs* file before the publicity meeting, Mr. Ryle," she said.

"Never heard of it," replied Ryle grumpily.

"It's on the top shelf. If I could stand on your chair—"

Ryle relinquished the chair. The girl got onto it, and reached up toward the row of dusty filing boxes ranged along the top of the bookshelf behind the desk. Taking one out, she opened it, withdrew a cardboard folder, from which, as she replaced the box with her other hand, an object slid out and fell onto the desk with a clatter, as though a thread that had held it invisibly suspended for days above Basil Ryle's head was at last frayed through.

On the desk lay a heavy cut-throat razor. It had sprung open, and where blade met handle there was a little red-brown spot. Basil Ryle stared at it. The secretary stared at him. A spasm, like a paralytic stroke, made one side of his face jerk convulsively.

"Oh, Mr. Ryle, whatever—?"

"Get on with it. Take the file to Mr. Geraldine. What are you waiting for?" Ryle's voice was calm, lifeless. "And this," he added, pointing to the razor, as the girl clambered down from the chair. "No! Not in your hand. Are you too highbrow to read detective novels? Here, wrap it in my handkerchief. Tell Mr. Geraldine where you found it, and not to mess it up with his fingerprints. Yes, blood does look like rust, doesn't it? Never mind."

Ryle's unnatural composure broke, and he went into a storm of laughter which drove the terrified girl out of the room. . . .

So much Nigel was told when he arrived. Arthur Geraldine's secretary, having given her account of it, was dismissed. The senior partner and Liz Wenham glanced covertly at Nigel; there was a look in his pale blue eyes which made them uncomfortable; they would have been far more uncomfortable if they had known what a cold, furious anger was mounting behind those eyes.

"Who asked for this file?" he curtly demanded.

"We're to discuss the publicity for a forthcoming volume of memoirs this morning," said Geraldine, "and we wanted to refresh our memory about the promotion campaign we did some years ago for a somewhat similar title."

"Yes, yes. But whose idea was it to look up the *Aston Memoirs* file?"

"Well, actually I think Miss Wenham suggested—"

"Have you rung the police?"

"No. I thought it best—"

"Then do so now, at once."

While Geraldine telephoned, Nigel opened the razor within the handkerchief, and exposed it to view on the senior partner's desk.

"An old-fashioned article, isn't it, Miss Wenham? And all complete—even to a bloodstain. Most instructive. So now Wenham & Geraldine can live happily ever afterward."

Liz Wenham's clear gray eyes wavered like pebbles in a stream. Her face had the clenched, stubborn expression of a child caught out in some misdemeanor.

"The Inspector is on his way," announced Geraldine.

"Let the dead bury their dead. Who's with Ryle? Protheroe?"

"Really, Mr. Strangeways!" Liz protested. "Did you expect us to put him under house arrest?"

Nigel's anger blazed out at last. "Good God in heaven! Who's talking about arrest? D'you mean to tell me you two have just left him to brood over this alone? When you know the state of mind he's been in? Are you incapable of thinking about anything but your net profits?"

Snatching up the razor in the handkerchief, Nigel ran along the passage to Basil Ryle's room.

The young man was sitting at his desk, so white and still that he might well have been a corpse. As Nigel entered, he raised his hands, which lay before him crossed at the wrists, as though handcuffed, then let them fall again on the desk.

"I thought you were the police. They're a long time coming."

"Snap out of it, Ryle."

"I must have done it. Don't you see?—this proves it." Ryle's chin was on his breast again; his voice was flattened out with utter dejection. Nigel took him by the shoulders, and shaking him hard, forced him to look up.

"It proves damn all. Here, look at the bloody great thing! No, *look* at it!" Nigel opened the handkerchief, laying the razor on the desk. "Do you realize what a police search is?

They searched the whole building a week ago. Do you really imagine they'd miss a thing like this?"

"You mean—?"

"I mean this razor was put into the filing box after the police search. It was put there yesterday or on Saturday afternoon."

"By the murderer?" asked Ryle, staring, visibly coming to life again.

"It is *not* the weapon the murderer used."

"But the blood—?"

"Anyone can draw blood. It's a commodity we all possess."

"Well, it beats me then. Whose razor is it, for God's sake?"

"I don't know for certain. But I think we shall find out that it belongs to a dead man."

XVI *FINAL PROOF*

"**O**h yes," said Nigel, "Cyprian Gleed has been arrested all right. This is by way of being a little private experiment of my own. What they call a 'reconstruction of the crime.' Just to work out the timing."

Nigel glanced once again at his wrist watch. Stephen Protheroe and Basil Ryle exchanged the simpering, embarrassed looks of members of an audience who have been called onto the platform to assist a conjuror. The three men were in Protheroe's room at Wenham & Geraldine's; it was late on Monday afternoon.

"The staff leaves the office in two waves," said Nigel, "the first at 5 P.M., the second at 5:30. There's ten minutes to go before the second exodus. Let's start."

He put on his macintosh, picked up a stout grip bag which lay on the floor, and led the other two out of the room. They took the lift to the ground floor.

"The side door is bolted till 5:30. Right?"

"Right," said Stephen.

Nigel went through the swing door and unbolted the side door beyond it, which gave onto the street. Then, followed by his two companions, he turned back and walked through the reception room and out by the main door, watched curiously by Miriam Sanders.

"I am the murderer," he said. "Observe closely everything I do, and don't ask unnecessary questions."

Taking out a key, he unlocked the side door. In the space

199

between it and the swing door, he whisked from his bag a duffel coat and a large black hat, put them on, peered through the swing door, then dashed for the lift, which was only a few paces away.

"You'll have to imagine the beard," he said.

"But how did he know the lift would be at the ground floor?" Ryle asked.

"Took a chance on it. Or maybe propped the door open."

"But—"

"Here we are. Follow me, don't get in the way, and remember that Miss Miles was expecting a visitor."

Nigel hurried down the passage, past the door of Stephen's room to the next one. They had met nobody so far. As he entered Millicent Miles' room, one of his companions gave a sharp exhalation, like the sound of a cat spitting; for the light was on in the room, and a woman sat at the table, typing, her back to the door.

"For God's sake—" exclaimed Basil Ryle.

"Shut up! I turn the key, behind my back. I put down the bag. Miss Miles doesn't look round, she's expecting me. I take one long stride."

Whipping a handkerchief from his pocket, Nigel stuffed it into the mouth of the woman at the typewriter, stifling her cry of surprise, and in the same movement tilted her back-ward, chair and all, so that her head was flung back and her throat exposed. Her feet began kicking at the desk; he dragged the chair away from it, took out an imaginary weapon, and passed it across her throat.

"I should be wearing gloves, of course," he said. "You'll observe that, holding her in this position from behind, I have her quite helpless. Also, I avoid the blood spurting out, except for my left arm which is clamped round her arms and body."

Stephen Protheroe shuddered violently. Nigel's level, ex-pository tones were like those of a surgeon giving his internes a running commentary upon an operation. His back pressed to the wall, as if he were trying to force himself through it, Basil Ryle glared at the scene.

"Only ten seconds or so since I entered. The victim is dying. The police reconstruction showed that it was done just like this. You must imagine the blood on my forearm and on the floor as I move to the next stage."

Standing a little away from the chair, Nigel let it down, with its burden, till its back was on the floor. He took the woman under the armpits, drew her away from the chair and dragged her to a corner of the room, where she lay with her black hair streaming over the dusty floor.

"I should be wearing galoshes," said Nigel. "The murderer put them on, over his shoes, at some stage before entering the room—in the lift, possibly."

Taking a staple from his pocket, and an ebony ruler from the desk, Nigel secured the sliding window.

"The door is locked. The window is fastened shut. Now I remove my bloodstained gloves and put on a fresh pair." Nigel went through the motions; then, setting the chair upright again, sat down at the typewriter.

"There's blood on the chair, but the duffel coat is expend-able—and anyway, I'm in a hurry."

Leaning over him, Stephen murmured, "Is all this grue-some detail necessary? I don't think Basil can stand it much longer."

Without replying, Nigel flicked the sheet of paper out of the machine, inserted another, and began typing.

"I'm copying from the actual sheet the murderer typed to insert in his victim's autobiography. This is where my timing may go a little wrong. I don't know how fast he typed, but presumably he knew by heart what had to be written. We shall get a time check in a few minutes, when the 5:30 exodus begins. Susan heard a typewriter in here as she passed by the door."

Nigel's dispassionate, almost pedantic tones heightened the unreality of the scene—an unreality whose focal point was the body of the woman sprawled upon her back in the corner, to which Stephen's eyes and Basil's kept reluctantly returning, Basil's in horror and Stephen's with a kind of puzzlement.

There was a confused noise of doors opening and footsteps passing. It was 5:30 P.M.

"An unnerving moment for the murderer," Nigel commented. "But that night nearly everyone on this floor had left the office earlier. . . . Right. I've finished typing this sheet. I substitute it for the relevant page in the autobiography—I daresay we shall never know what that page contained, but we can be sure it gave a clue to the murderer's identity. O.K. But I'm still not quite happy. Some instinct tells me that my victim may unconsciously have put in another bit of evidence against me, since I last read the typescript. I leaf back through the sheets—I need only worry about the chapter describing a certain period of her younger days. I see, in the margin, a penciled capital letter. It might be a G. or a C. It gives me quite a turn. I rub it out."

"'G'? Gleed?" asked Ryle in a dry, harsh whisper. "But she wouldn't—"

"It couldn't be Gleed, or Cyprian; he wasn't born till ten years later," said Nigel.

"Try 'G' for Geraldine," Stephen suggested.

Basil Ryle stared uncomprehendingly at him.

"I'm satisfied there are no more marginal references to me," Nigel resumed. "Now comes rather a nasty bit. And this is where the murderer made two mistakes."

Lifting the typewriter, he laid it on the floor beside the woman's body. He took up her flaccid fingers, wiped each of them, then, after polishing the keys of the machine, pressed her fingers upon them.

"You see what's wrong?" He looked up sharply at Ryle. "Well—no."

"She was a touch typist. I'm putting her fingerprints on the wrong keys. And now I make my second mistake. I replace the typewriter on the table, so. I insert the sheet my victim was typing when I surprised her. But I don't get it back in alignment. That's why the police first suspected the murderer had done some typing. A fatal oversight on his part."

Nigel proceeded to remove the staple which held the sliding window fast. After a slow look round the room, he said,

"Well, that's all for here. What's the time? Five forty-two. We're running a little late—the murderer was seen leaving the building at 5:40. Come along."

"But—"

"Oh, yes, galoshes. He probably removed them at the door and put them in the bag."

"But for Christ's sake!" exclaimed Basil Ryle, pointing at the woman lying in the corner, eyes closed, breathing equably, a tumble of black hair about her pale face. "*Who is that?*"

"Oh, she's my dummy."

"Won't you introduce us?" said Stephen.

"No time. Come along." At the door, Nigel said, "Thank you very much, Clare. That will be all. See you later."

They went down in the lift. As they passed through the swing door, Nigel said, "The murderer was seen at this point, wearing black hat and duffel coat. No doubt he removed them and put them in his bag after the swing door closed behind him."

In the confined space, Nigel jostled the other two as he took his macintosh from the bag and exchanged it for the duffel coat. Stephen Protheroe looked worried but interested; Basil Ryle seemd to be in a daze. They went out into Angel Street, joining the stream of city workers which flowed toward Embankment Gardens like a tributary winding to the Thames. They passed through Charing Cross Underground Station and climbed the steps up to the northern end of Hungerford Bridge. Nigel's companions had some difficulty in keeping up with his long stride; he moved in an abstraction which they found formidable. Basil Ryle, however, broke the silence:

"Do you mean to say that Cyprian Gleed just walked into the office, in his usual clothes, and—and did all that?"

" 'Usual clothes'? Would you expect the murderer to wear fancy dress?" Nigel harshly answered.

"I'd never have thought he'd have the nerve."

Nigel stopped abruptly—they were now halfway across the bridge—and gazed meditatively at his companions.

"Perfect hatred casteth out fear," he said.

Stephen Protheroe's fine eyes held Nigel's for a moment, as the three moved closer to the rail of the bridge to let the hurrying commuters pass.

"Why did you ask us to watch this reconstruction?" Stephen's fishlike mouth nibbled for words. "I mean, you could have worked out the timing by yourself."

"I thought you'd be interested," was Nigel's stony reply. He turned toward the river, with the arc lamps of the bridge behind him. Lights from the Embankment scribbled hieroglyphs on the restless water below. The Shell-Mex Building and the Festival Hall blazed at each other across the Thames. A train rumbled onto the bridge, with a long-drawn clattering reverberation. As it passed, Nigel released the grip bag which he had been holding over the rail, out of sight.

"You see? Nobody noticed anything."

"Noticed what?" Basil asked, almost shouting.

"You didn't even notice yourself. A murderer dropping into the Thames a weighted bag containing a duffel coat, a large black hat, some gloves, a pair of galoshes, a razor, and one thing more." Nigel gradually lowered his voice, as the thunder of the electric train receded.

"One thing more?" Ryle shook his head as if to clear it of bewilderment.

Nigel turned to Stephen. "Can't you tell him?"

There was a little silence, then Stephen said with a frown, "But you told us that Cyprian Gleed had been arrested."

"I never said he'd been arrested for murder."

"What *is* all this?" Ryle edgily asked. "I can't make head or tail of—"

"In that case"—Stephen Protheroe's resonant voice was clearly audible through the shuffle and clatter of the passers-by—"the 'one thing more' could have been a false beard."

"Correct."

"Then the murderer disguised himself to look like Cyprian Gleed?" Basil's face changed from consternation to a kind of timid hopefulness.

"Come on. We're wasting time." Nigel led them past the Festival Hall. Four minutes later they were passing through the main booking office of Waterloo Station. The clock hands stood at 5:57, as the three men emerged into the great space filled with a rush-hour crowd hurrying toward the departure platforms.

"You see," said Nigel, "in spite of my lecture en route, there's still time."

"Time for what?" Stephen Protheroe's face was working like a disturbed anthill.

Nigel looked full into his eyes, saying sadly, almost remorsefully, "Time for you to pick up your bag—your *weekend* bag—at the Left Luggage Office and catch the 6:05."

Involuntarily, Stephen's eyes switched toward the Left Luggage Office. Inspector Wright and Detective Sergeant Fenton were standing at the withdrawal counter. Stephen gave a little nightmare whimper, then darted away, running diagonally left toward the open gate of a departure platform which was sucking in its last few belated travelers like a drain.

Nigel started after him. Wright and the Sergeant moved to cut him off. But to get through the solid stream of travelers hurrying from right to left across Nigel's path was like swimming at right angles to an irresistible current. Bumping and tripping, he was carried off course and lost sight of the quarry he could not, in any case, quite wholeheartedly pursue. When he set eyes on Stephen again, the little man was not where he had expected him to be. Under cover of the crowds, he had doubled sharply right like a coursed hare, and was now at the barrier of an arrival platform some thirty yards away.

Nigel waved frantically to Inspector Wright, who came breasting up the stream of passengers toward him. He turned again, making for the barrier, to see—with a stab of anguished compunction—that Stephen Protheroe was getting a platform ticket from the slot machine there. At the same moment he became aware of a sound like the first drum roll of distant thunder, a vibration in the air which was felt rather than heard above all the confused din of the station.

"No! Don't do it! Don't!" Nigel silently cried. The ticket collector, still oblivious to Wright's shouts and waving, punched Stephen Protheroe's ticket. The thunderous vibration swelled. Side by side now, Wright and Nigel tore through the gate. Twenty yards ahead of them, running away up the platform, Stephen looked round once, dodged a porter who tried to intercept him, and raced on to meet the great locomotive which drew its train, gradual as lava, toward the buffers—raced to meet it, as if it were his salvation, not his doom, and threw himself under the iron wheels.

"When did you first suspect it was Stephen?" Basil Ryle asked.

" 'When did you first begin to love me?' Who can answer that?" There were exhaustion and bitterness in Nigel's voice. Clare put her hand over his. The three were in Clare's studio, late that same night. Nigel turned to Ryle: "Sorry. I'm feeling foul. I liked him."

"Let's skip it, then." The color was back in Ryle's cheeks; he had been looking round the studio, relishing like a convalescent after a dangerous illness the mere existence of everyday objects. Nigel began to answer his original question.

"It was an entry in Miss Miles' diary for the day she was murdered. 'Thorsday?!' What could that mean except a showdown about General Thoresby's book? Cyprian Gleed told me he'd seen his mother, alone in Stephen's room, bending over the proof copy, the morning it went back to the printers. If it was she who had tampered with it, and Stephen who had presented an ultimatum, she'd never have written that word in her diary, and with an exclamation mark. Therefore, the position was vice versa. Even the pun told its story. She hated puns. Stephen made them, just to annoy her. So, in her childishly vindictive way, she made a pun of the time when her ultimatum was due to expire."

"But the motive seems so feeble," said Ryle. "Do you mean to say he killed her just because she threatened to expose him to the partners for mucking about with that proof?"

Clare said, "It'd only have been her word against his, anyway. Why should the partners have believed her?"

"His job, his security were at stake. So he thought. It was her threat that lit the train. But the train itself—a long history of brooding hatred—it hardly bears thinking about." Nigel broke off for a little. "What you don't realize is that Stephen had been living for months with nothing but a thin wall and a sliding window between himself and the woman who had drained the meaning out of his life. Their child—"

"What's this? Whose child?" Ryle exclaimed.

Nigel told him the story of Paul Protheroe. "Stephen inserted that fake page in the autobiography because the original page would have shown us a connection between him and Millicent, shown they had had a child, and given us his reason for tampering with the proof—as soon as we'd discovered that a Paul Protheroe was killed at Ulombo."

"I thought you believed his story that his brother had been the child's father," said Clare.

"I did at first." Nigel explained to Ryle where Peter Protheroe came in. "You see, Stephen had two lines of defense. The first was to prevent us finding out about his son at all. When I broke through this, he fell back upon the second. It was at lunch in his flat. He then said that Millicent had had the child by his brother, Peter, a student for Holy Orders, and that he himself had taken the responsibility for it to prevent Peter's career being ruined at the start. It was ingeniously done. The new story couldn't be disproved, since Peter had died a year or two later in the mission field."

"What made you doubt it?"

"Stephen's poem. And his failure to write any more."

"I don't get it," said Ryle.

Rising to his feet, Nigel prowled restlessly about the studio. "*Fire and Ash* is the key to the whole wretched business," he said at last. "Stephen told me the Peter tale with extraordinary conviction. He made me *feel* the horror of the ordeal, from seduction to heartless repudiation, through which Millicent had dragged an idealistic young man. *Fire and Ash* was written, he said, out of his brother's experience, which he himself had observed. All right. But then I discovered he had tried later, again and again, to write poems, and they were all failures. If *Fire and Ash* had been the product of mere secondhand experience, why on earth should Stephen's faculty of imagination and poetic sympathy have utterly failed him over the subjects he attempted afterward? The fact he did so fail made me more and more certain that *Fire and Ash* must have come from firsthand experience—from a personal agony which had blasted his talent as well as disrupting his life. Poets are very tough inside, you know. They don't give up the ghost because of anyone else's sufferings. But occasionally a poet's talent is killed by some traumatic experience of his own. I believe that, after what he went through with Millicent as a young man, he determined never to be vulnerable again; he retired into himself, scorching the earth behind him—and the poetic seed got burnt up in the process."

"Yet he found himself vulnerable to Millicent still, after all those years?" asked Clare, musing.

"Yes. His son, Paul, was the one thing he had saved from the wreck. Poetry failing him, he had transferred all his hopes and aspirations to the boy—centered all his affection upon him. Then Paul was killed. Stephen had nothing left. You can see how irresistible was the temptation to expose the Governor whose bungling had caused the boy's death."

"But Millicent—?" Ryle began.

Nigel turned full toward him. "Doesn't it occur to you that one of Stephen's motives for killing her was to prevent her destroying you? Prevent her making the havoc of your life she had made, thirty years ago, of his own?"

"Oh lord, oh lord," muttered Ryle, unable to meet Nigel's eyes.

"One motive—a minor one, though. Then there was his literary integrity: he would not be blackmailed into withdrawing his opposition to the firm's publishing her trashy novels."

Basil Ryle flushed to the roots of his red hair.

"But that was a subsidiary motive, too," Nigel continued. "There he was, living for months cheek by jowl with a woman who had struck one great poem out of him and finished him in the process. Oh, the wound began to bleed again, all right. I've no doubt she opened it wider and wider. I heard her say to him once, 'I do *write*, anyway.' She had plenty of opportunity to taunt him privately, through that sliding window. Jean heard her say to him, 'You're just impotent.' I'm sorry, Basil, but she was a bird of prey. She'd peck at him till he was a tangle of raw nerves, just as she did thirty years before, and a good deal more skillfully."

"Yes. She was skillful enough at that." Basil Ryle's voice was almost inaudible.

"She'd used her pregnancy once as a weapon against him, and I dare say he'd been near killing her then. Now she had another weapon, another way of exercising power over him— two stet marks on a proof copy. And this time he did kill her. She'd asked for it several times too often."

"I see all that," said Ryle presently. "But I'm surprised he tried to involve Cyprian Gleed. It doesn't seem in character."

"It may have been an improvisation on his original plan. Not that he would have felt much compunction about it, when he compared that worthless young man with Millicent's other son—his own. Anyway, he'd planned to kill her on a Friday, when the staff would all have left the office soon after 5:30. He'd fixed up to stay the weekend with friends in Hampshire; he had a standing invitation to visit them; yet, as he himself told me, he seldom went out of London; and he'd never stayed with them before. That alone should have roused our suspicion—in a murder investigation one always looks for the unhabitual, the non-routine action. He'd worked out a fairly neat alibi—impressive without being blatantly watertight. Then, the day before, he overheard Millicent talking to her son on the telephone: she was going along to Cyprian's flat, after she'd had her showdown with Stephen the next evening: Cyprian would be awaiting her there, alone. Cyprian would have no alibi for the time of the murder. This may have given him the idea of dressing up as Cyprian; or it may have been part of his plan all along—we'll never know. Stephen had already pinched the spare key of the side door—in order to confuse the issue generally, I should think: but its disappearance would point the finger at Cyprian in particular. Stephen bought a duffel coat and a large black hat. Unfortunately he couldn't grow a beard overnight, so he had to buy a false one."

"Unfortunately?"

"Susan, when she described the man she'd seen leaving the office, used an odd form of words. 'He'd got sort of hair at the side of his face,' she said—she only saw his back view, remember. I was alerted by that. How peculiar that she shouldn't just say, 'He had a beard.' Unconsciously she had registered it as a pseudo-beard. So the idea came to me of the murderer disguising himself as Cyprian Gleed just in case by ill luck he met anyone on his way to or from Miss Miles' room. Why should he not have taken off the bloodstained duffel coat *before* leaving her room? Because, if he were to be seen, he had to be seen wearing it. The one suspect who *couldn't* afford to be recognized in the office at that time was

Stephen, who had been seen by Miriam Sanders leaving the building at 5:20 with a "weekend bag." He had to do the murder when he did because (a) Millicent would not wait for him much longer to keep his appointment with her, and (b) because his alibi partly depended upon Miss Sanders' seeing him go out of the building; and she herself would be leaving it at 5:30. Mind you, he never expected the time of the murder to be fixed so accurately by the police."

"How did they fix it?"

"By spotting that the sliding window had been nailed fast. He didn't anticipate their noticing it. When the police first questioned him, he made the ghastly mistake of telling them that he knew nothing of the window having ever been nailed up. If he'd said it had been, and given some plausible reason for it, Wright might well have thought no more of it. However. It could only have been nailed up as a precaution against someone popping into Stephen's room and looking through into Millicent's. Therefore the murder must have been done while there were still people in the office—and even the partners never stayed much later than 6 P.M. on a Friday. Therefore the time of the murder could reasonably be narrowed down to between 5:20 and 6."

Clare gave Basil Ryle some more brandy. Ever since coming to the studio, he had eyed her from time to time in an incredulous way, as though he found it difficult to identify her with the dummy of Nigel's macabre reconstruction. Now he said,

"I'm afraid I never took to Stephen very much. Couldn't discover what made him tick. And of course he was one of the old guard, while I was in the firm on sufferance—a brash little clever scholarship boy. But I'd never have thought he'd try to incriminate me."

"He didn't."

"But that razor—"

"The razor was planted in your room by Liz Wenham."

"*Liz?* Oh, come off it!"

"Liz had a very soft spot for Stephen Protheroe—that was obvious to me from the start. She's an extremely intelligent woman, so she had no difficulty in perceiving the strength of the case I was building against Stephen, or in picking up the hints I let fall to her about it. Moreover, her love for him had made her instinctively aware of his hatred for Millicent. I believe she suspected him first during that dinner party in the Geraldines' flat."

"How so?"

"You remember, I mentioned the telephone call Gleed had made to his mother the day before the murder—the one fixing an appointment with her at his flat. Stephen said he 'couldn't have heard what she said.' I caught a curious, apprehensive look on Liz's face just after. My guess is that she'd been in Stephen's room when that telephone call came through, and heard it because the sliding window was open; so why, she wondered, did Stephen say he had not heard the conversation? Anyway, when I interviewed her last, she could not conceal her worry about Stephen, and I deliberately dropped some clanging hints. On my way out, I overheard her ringing him up, asking him to lunch. Heaven knows what passed between them—Stephen wouldn't consciously let anything out, I'm sure—but her suspicion was confirmed, and—"

"But why should she pick on *me?*" asked Ryle irritably.

"You've already answered that: because you were not one of the old guard—not a Wenham, or a Geraldine, or a Protheroe. And she loved Stephen, in her peculiar way. Anything to divert suspicion from him. Now, outside the job, Liz is a naïve, unworldly type. And impulsive. On an impulse to protect Stephen, she rummaged out that old-fashioned razor—belonged to her father, no doubt—her house is a museum of Wenham relics; she cut herself with it, then rushed off to plant it in your room at the office. It was she who asked for that file to be looked out on Monday morning."

"Well, I must say—" began Ryle in an aggrieved tone that sounded faintly absurd.

"No doubt she's very ashamed of herself now—of being so incompetent as to suppose that the police wouldn't have searched your room thoroughly enough to find it. But, as I say, she did it on the spur of the moment, panicking about Stephen. And it would fit in with the theory that you'd killed Miss Miles in a brainstorm. There's a lot of ruthlessness in Liz; she's a monomaniac, you must remember, about the firm's reputation; and the only other thing in life she was devoted to was Stephen."

Basil Ryle gave Nigel an almost inimical look. "While we're talking about ruthlessness—"

"Oh, Geraldine's the same in his way. But he's got a softer center. He'd been shaking in his shoes about an early episode with Miss Miles. There are two versions of it, and I don't give a damn which is the true one. He had to use Liz as an intermediary to convey his own version to me. He's a moral coward; but he had no motive conceivably strong enough for murder."

"I'm not talking about Geraldine. I'm talking about your demonstration this evening. Was it necessary to put Stephen through that ordeal? Damn it, man, it was absolutely sadistic. Surely the police would have got him sooner or later? Traced his purchases of the false beard and the duffel coat?"

"Yes, I expect they would," replied Nigel mildly, "once they started inquiring for Stephen in connection with them, instead of Cyprian Gleed."

"Well then, how can you justify that melodramatic business this evening? It was like—like torturing a man to soften him up for a confession."

"For God's sake, Mr. Ryle!" Clare's voice trembled with indignation. "Don't you even understand that? Do you think Nigel was enjoying himself?" She rose, with a swirl of black hair, and stood over Basil Ryle. "You've said it yourself—the police would have got him 'sooner or later.' But it had to be sooner. For your sake." Clare stamped her foot. "Your sake. Nigel was afraid for you—for what might happen to you if you were left in suspense any longer, wondering whether you hadn't done it in a brainstorm. That's why Stephen Protheroe had to be broken down quickly. You should be grateful to him, instead of taking up this sentimental attitude toward—"

"All right, all right. I'm sorry. Really. I never thought—" Basil Ryle offered Nigel his hand; then, with a glance at Clare and a glint of his old perkiness, added, "You're a lucky chap."